The Deep-sky Imaging Primer

Charles Bracken

The Deep-sky Imaging Primer

by Charles Bracken

Copyright © 2013 Charles Bracken

All rights reserved. No part of this publication may be reproduced, distributed, or transmitted in any form or by any means, including photocopying, recording, or other electronic or mechanical methods, without the prior written permission of the author, except in the case of brief quotations embodied in critical reviews and certain other noncommercial uses permitted by copyright law. Permission requests should be sent to the author at DeepSkyPrimer@gmail.com.

All trademarks used are the property of their respective owners.

Printed in the United States of America

ISBN-13: 978-1481804912

ISBN-10: 148180491X

First Edition

To my wife and children—thank you for patiently tolerating the many hours I spent creating this book.

Introduction

With electronic imaging technology, amateurs can now capture images of the night sky that reveal vistas unseen only a few decades ago. Astronomical imaging no longer requires an observatory or expensive scientific equipment. All you need are clear skies and a modest investment of time and money.

This book will guide you in creating beautiful images of the night sky. All of the essential concepts are laid out along with examples of how they work and what they mean to the final image. The material is divided into three sections:

> **Understanding Images** begins with the fundamentals of how electronic imaging works, with special attention given to the concepts of signal and noise. This will help you make informed decisions about equipment and processing.
>
> **Acquiring Images** explains how to use a telescope and camera to capture deep-sky images. Everything about equipment and technique is covered in this section.
>
> **Processing Images** covers the important work done at the computer to combine the raw images and process them into a masterpiece you'll be proud of.

I have tried to give practical advice using current equipment under the circumstances faced by the most amateur imagers. In fact all of the images in the book were taken under heavily light polluted suburban conditions forty miles from New York City.

Only in the past century have we come to realize the breathtaking scope of the universe. It is inspiring to create an image from light that left its source thousands, millions, or even billions of years ago. With each image, we understand a little more about the universe and ourselves. I hope this book will help you create the best images your skies and equipment can produce.

Clear skies to you!
—Charlie Bracken

Note: I am always interested in hearing your feedback and using it to improve and expand future editions. Please contact me at: DeepSkyPrimer@gmail.com to share your thoughts.

Contents

Introduction ... v

I. Understanding Images ... 13

1 **Electronic sensors** 14
    The challenges of astronomical imaging 14
    How electronic sensors work 14
    Well capacity, gain, and dynamic range 16
    The importance of bit depth 18
    Response curves and raw files 18
    Creating color images 20
    Sensor architecture 21

2 **Signal and Noise** 23
    Signal and noise at the pixel level 23
    Shot Noise 24
    An example of shot noise 26
    Skyglow 29
    Thermal signal, hot pixels, and dark frames 30
    Read noise and quantization noise 32
    Two-dimensional corrections 33
    Sensor non-uniformity 33
    Uneven illumination 33
    Adding signals and noise 34

II. Aquiring Images ... 37

3 **Mounts and alignment** 38
    Choosing and using a GEM 39
    Meridian flip 39
    Polar alignment 40
    Tracking error 41
    Drift alignment 42

4 **Cameras** 43
    DSLRs 43
    Dedicated astronomical cameras 43
    Sensor size and pixel size 44
    Connecting a camera to a telescope 44

5 **Optics** 46
    Telescopes for visual vs. imaging use 47

    Optical aberrations 47
    Refractors 48
    Telephoto lenses 50
    Reflectors and compound telescopes 51
    Telescope quality 54

6 Image scale: matching sensor, and optics 55
    Resolution and seeing 55
    Sampling 56
    Oversampling revisited 58
    Field of view 59
    Equipment recommendations 60
    Focal reducers and field flatteners 61

7 Choosing appropriate objects to image 64
    Getting a sense of scale 64
    The deep-sky catalogs 65
    A survey of object sizes 66
    Giant Objects (2° or greater) 66
    Really Big Objects (1–2°) 67
    Objects from 30-60' 67
    Objects from 15-30' 67
    Objects from 5-15' 67
    Objects from 1-5' 68
    Objects smaller than 1' 68

8 Focusing and autoguiding 69
    Focusing 69
    Achieving critical focus 70
    Autoguiding fundamentals 71
    Flexure 72
    Connecting computer to mount 72
    Software and settings 73
    Choosing a scope and camera for autoguiding 75

9 Setup and accessories 76
    Reducing setup time 76
    Other important equipment 76
    Dovetails, rings, and other mount accessories 77
    Power supplies 77
    Dew prevention 78
    USB hubs and extension cables 78

10 Filters and narrowband imaging 79
    Filters for imaging 79
    Light pollution filters 82

11 Taking the exposures 83
    Controlling the camera 83
    Choosing exposure duration and gain 84
    Planning for a night of imaging 85
    An image capture workflow 86
    Dark frames 87

          Flat frames   88
          Bias frames   89
          Dithering light frames   89

12   Atmospheric effects   90
          Light pollution   90
          Target altitude   91
          Local turbulence   93

13   Diagnosing problems and improving image quality   94
          Wind or tracking errors   94
          Mirror flop   94
          Diffraction patterns   94
          Focus   95
          Halos   95
          Tips for better image capture   96

III. Processing Images   99

14   Color in digital images   100
          File formats   100
          Visual response to color   100
          Producing color on print and screen   101
          Color management and color spaces   103
          LAB color   103
          The HSL/HSV/HSB color model   104
          DSLR white balance   104
          Deep-sky color accuracy   105
          Color calibration with G2V stars   105

15   The calibration process   107
          Calibration exposures   107
          Stacking parameters   110
          An example of calibration and stacking with DeepSkyStacker   111
          Throwing out problem shots   114
          Aligning monochrome images for color combination later   114
          The drizzle algorithm   115
          Diagnosing defects in calibration output   116

16   Principals and tools of post-processing   118
          Selective adjustments   118
          Understanding the histogram   118
          The curves tool   121
          Levels   123
          The magic of layers   124
          Layer blending modes   125
          Layer masks   126
          Creating a composite layer   126
          Selections and feathering   127
          A post-processing workflow   128

17 Stretching: reallocating the dynamic range 129
Non-linear stretching 129
Start with balanced colors 130
Initial development 130
Object-background separation curves 131
Intra-object contrast curves 131
Controlling highlights and boosting saturation 133
Posterization: the perils of over-stretching 133
The effect of stretching on color 134
Digital Development Process (DDP) 135

18 Background adjustments and cosmetic repairs 136
Gradient removal 136
Repairing satellite trails, dust spots, and reflections 138
Correcting elongated stars 139

19 Color synthesis and adjustment 140
RGB Color Synthesis 140
Creating a luminosity channel 140
Narrowband RGB synthesis 141
Further hue adjustments 142
Using clipping layer masks to map color 143
Blending RGB and narrowband data 143
Color balance 144
Enhancing color saturation in RGB 144
Boosting saturation in LAB 145

20 Sharpening and local contrast enhancement 147
Convolution, deconvolution, and sharpening 147
Unsharp mask and local contrast enhancement 148
Selective application with layer masks 151
The high pass filter 151
Other ways to create local contrast 153

21 Star adjustments 155
Star removal and star selection 155
Reducing star size across the image 158
Reducing the brightest stars 159
Correcting star color in narrowband images 160

22 Noise reduction 162
Visual noise, color, and scale 162
Tools for noise reduction 162
Applying noise reduction selectively 163

23 Composition 165
Framing the scene 165
Orientation 165
Color 166
Walk away 166

24 DSLR Processing example: The Witch's Broom Nebula 168

25 CCD Processing example: The Rosette Nebula in narrowband   175

A. Exercise Answers   185

B. Moonless hours   189

Index   195

1 Understanding Images

This first section briefly covers what's going on "under the hood" in the electronic imaging process. Understanding how electronic images are created is uniquely important to long-exposure astronomical images. If you are math-phobic, don't worry: the concepts are more important than the formulas. These ideas form the foundation of quality images, and we'll come back to them again and again later as we explore equipment, image capture, and image processing.

M31, The Andromeda Galaxy
(18 10-minute luminance exposures combined with approximately 11 hours of color exposures)

1 ELECTRONIC SENSORS

The challenges of astronomical imaging

Deep-sky objects include galaxies, nebulae, and star clusters. Many are stunningly beautiful, and they range in size from tiny planetary nebulae and distant galaxies only arcseconds across to gigantic nebulae larger than the full moon. All of them, however, are dim, and this is the main challenge of creating deep-sky images.

When faced with a dimly lit outdoor scene, a daylight photographer will use a faster focal ratio or take a longer exposure. In our case, we'll do both, but faster optics can only take us so far. The defining characteristic of deep-sky imaging is that we must take very long exposures. Instead of measuring exposure times in fractions of a second, we will measure them in minutes. Further, we'll synthesize many images together into one that incorporates hours of total exposure time.

None of this was possible before electronic sensors and digital imaging revolutionized photography. The quality of deep-sky images has taken a quantum leap forward as a result, allowing amateurs to create images that were impossible in the film era.

The process is not easy, however, as there are many challenges to overcome. Aside from being very dim, the sky rotates slowly over our heads, so we'll require accurate tracking. We'll also see that while daytime photography is relatively forgiving when it comes to optics, the night sky (especially stars) seems designed to reveal even the slightest optical aberrations. Skyglow fills the sky for most observers anywhere near civilization, obscuring our incredibly dim targets. And while most photographers give only passing thought to concepts like signal and noise, we'll see that they play a crucial role in nearly every aspect of deep-sky imaging.

Fortunately, there are terrific tools and techniques within the reach of amateurs to overcome every one of these challenges, and we'll cover all of them on our journey. But let's start with the tool that makes all of this possible: the sensor.

> Planets are very small, with all of their light concentrated into a nearly-stellar point. Thus planetary imaging requires short exposure times at very high magnification. While both disciplines have some similarities, the equipment and processing techniques are quite different. This book focuses on deep-sky imaging.

How electronic sensors work

Photons are the fundamental particles that carry light. The job of the electronic sensors in cameras is to count photons, and their sensitivity is amazing. They can register a signal discriminate between brightness levels resulting from only a few photons. The individual sensor elements on a sensor are referred to as photosites, each of which corresponds to a pixel in the final image. "Pixel" is short for picture element, and following the same logic, photosites are sometimes also called sensels. Colloquially, most people ignore the distinction between a photosite and the resulting image pixel—a point on the image and a point on the sensor both commonly referred to as a pixel.

Figure 1. A typical CCD sensor

There are several steps between photons striking the camera's sensor and the image file on your computer. Each step is an opportunity for inaccuracy, known as noise in this context, to be introduced. Photons pass through your optical system then strike the silicon of your sensor. This creates a tiny amount of electrical charge in the form of electrons in the photosite 'wells.' Not every photon that strikes the sensor creates a charge. The silicon will reflect some wavelengths of light, and even photons of the right wavelength do not always generate a charge. The efficiency with which a sensor can convert photons into electrical charge is called its quantum efficiency. When the exposure is over, circuitry in the camera called an Analog-to-Digital Converter (ADC) quantifies the charge at each photosite, creating a digital value that is proportional to the number of photons that struck the photosite. The units for this digital value are sometimes referred to as Analog-to-Digital

Units (ADUs). Finally, after a value has been assigned to every photosite, the camera can arrange all of the data into a computer-readable file. The steps of the imaging process are shown in Figure 2.

Understanding Images

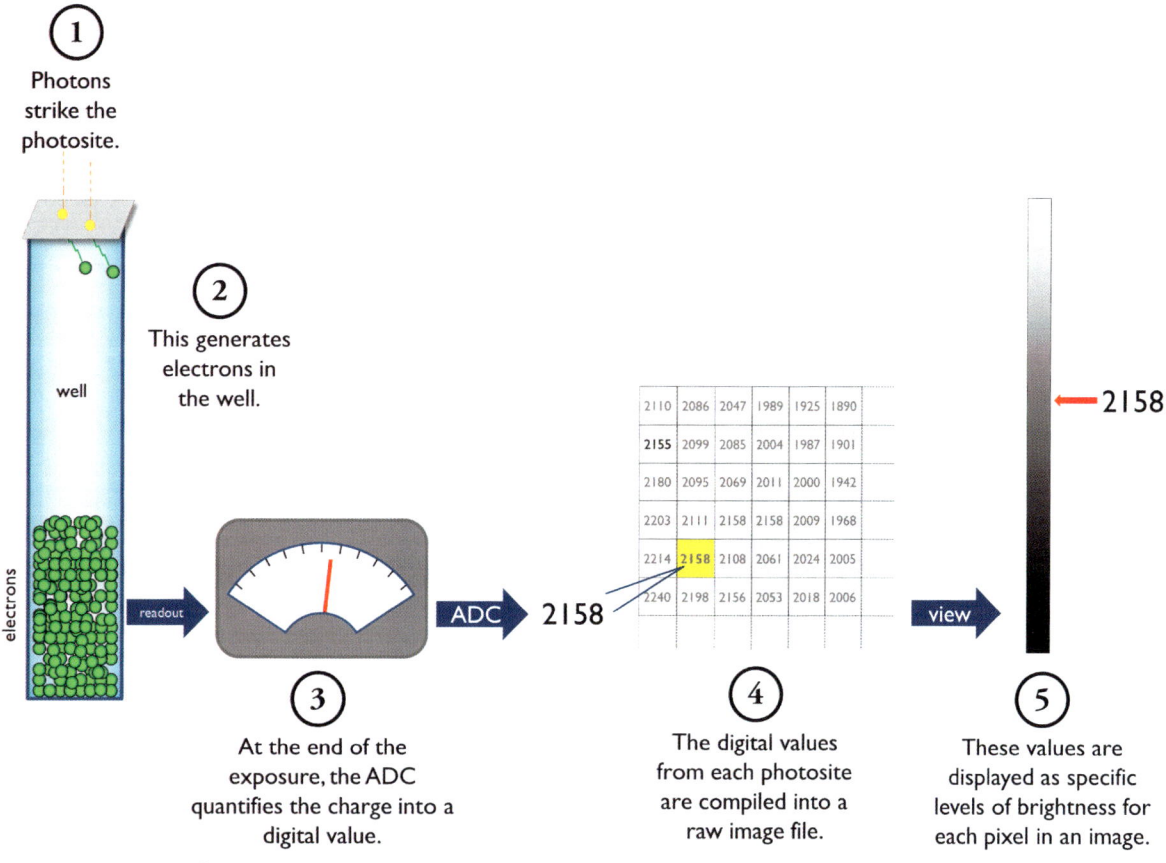

Figure 2. How electronic images are created

While CCD and CMOS sensors are slightly different sensor technologies, especially in the way the cells are "read out," for most purposes they are interchangeable. In the past CCDs had better noise and efficiency profiles than CMOSs, but improvements in CMOS chips have eliminated any differences from the user's perspective. The details of how CMOS and CCD sensors differ are not generally of any consequence to most imagers, as the quality of results is equal. Both technologies convert photons into an electrical charge, thus creating a signal that is proportional to the brightness of the light focused on a given photosite. Both technologies also discharge the sensor in order to read out the image, so there is no way to measure the image during the exposure.

Because of the inherent properties of silicon, sensors only respond to photons in the approximate wavelength range of visible light. Infrared or longer wavelength photons do not carry enough energy to free an electron to generate a charge, and those in the ultraviolet range are reflected. The spectral response of a sensor can be altered adding other elements or compounds to the silicon, but in general, the response range for most sensors is approximately 300 nm to 1,000 nm. By comparison, the human eye can see wavelengths in the range of about 400 to 700 nm.

Compared to film, digital imaging for astronomy is a huge leap forward. The best black and white films have a peak quantum efficiency of around 1–3%, while most commercial CCD and CMOS cameras are in the 30–60% range. The best scientific CCD cameras have efficiencies of over 90% in some wavelengths. These differences alone dramatically improve performance in capturing low-light images, but the fact that the image is stored as digital data rather than a chemical pattern on an emulsion has even more profound consequences. Many individual digital exposures can be combined in a processed called "stacking" to yield one image containing hours of data with noise characteristics nearly equivalent to a single long exposure. Even better, powerful image processing software and techniques are available to pull every bit of detail from an image and allow adjustments to specific areas. All of these factors have brought incredible deep sky images within the reach of amateurs using consumer grade electronics.

It all starts with photons at the sensor, and quality images begin with quality data. No amount of processing can make up for poor raw data, so let's start with the sensor and how its attributes will affect the final image.

Well capacity, gain, and dynamic range

Each photosite can capture only a certain number of photons before it becomes full of electrons and ceases to register further strikes. This limit is called the full well capacity. Typical full well capacities range from 20,000 to 200,000 electrons, and the capacity is related to the size of the photosites. The size of a photosite imposes a physical limitation on capacity, so the maximum well capacity is proportional to the area of each photosite. A rule of thumb is that there are about 1,000 electrons of well capacity per square micron. For example, in Kodak's KAF line of CCD sensors, the well capacity of the KAF-8300 with 5.4 μm sites is about 25,000 electrons, while the KAF-1001E with its giant 20 μm sites holds over 200,000.

When the analog-to-digital converter reads each photosite, it maps the voltage it senses into a digital value from a predetermined range called the bit depth. 8-bits of depth means there are 256 possible values (two to the eighth power) that can be assigned, corresponding to the range of values expressed by eight binary digits (from 00000000 to 11111111). In fact "bit" is short for "**b**inary dig**it**." No voltage is recorded as 0, and the maximum voltage the photosite can hold (a "full well") is recorded as 255. 12-bits cover a range of 0–4,095 and 16-bits from 0–65,535. Most DSLRs have ADCs that generate 12- or 14-bit data, while most recent dedicated astronomical cameras have 16-bit ADCs. As you can see, having an ADC with a greater bit depth means that it can distinguish smaller voltage differences, yielding a more accurate assessment of the brightness of each pixel.

The bit depth of the ADC is not the end of the story for accuracy, though. Being able to resolve more discrete levels of brightness is meaningless if this resolution is overwhelmed by noise. For instance, while cameras with 14-bit ADCs produce a potential range of values that is four times larger than 12-bit ADCs, tests of current DSLRs show that for most of them, those additional two bits do not offer any additional precision. The inaccuracies of reading the photosites (read noise) are larger than the detail supposedly offered by those two bits. Dedicated astronomical CCDs are more likely to realize a true benefit from 14- or 16-bit ADCs due to their higher well capacities and lower read noise.

Some cameras, including all DSLRs, have adjustable gain settings that allow the voltage captured in the photosite wells to be multiplied. This increases the sensitivity of the camera. The ISO setting on a consumer digital camera mimics the sensitivity levels of film, and this is accomplished through applying different amounts of gain to the output of the sensor before the ADC quantifies the charge. Some astronomical cameras also allow gain adjustments that work the same way. Increasing the sensitivity, however, occurs at the expense of dynamic range.

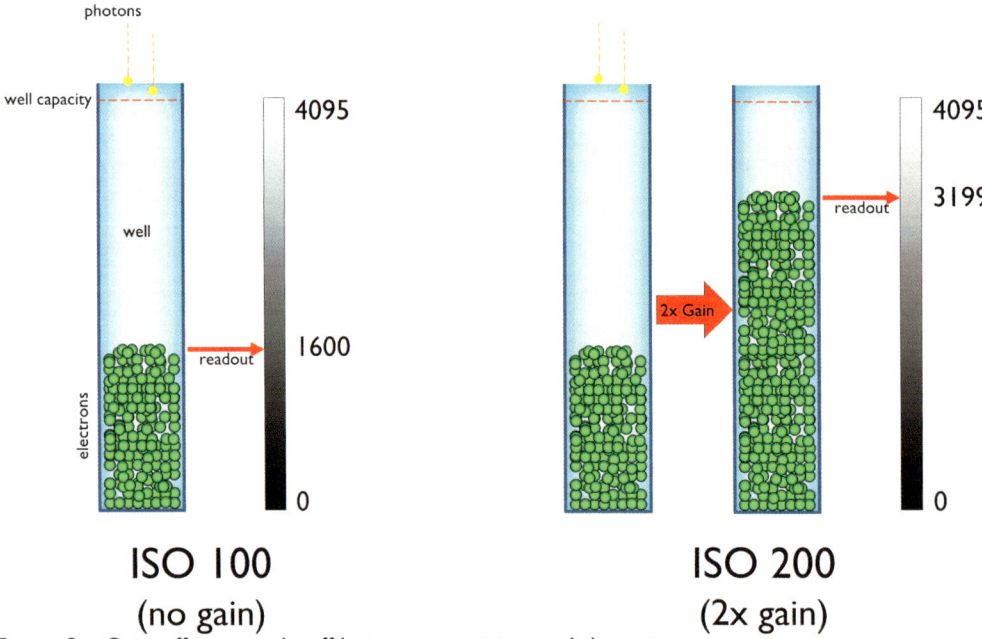

Figure 3. Gain offers a trade-off between precision and dynamic range

Dynamic range is the ratio of the brightest and dimmest signals. There are two dynamic ranges to consider: that of the scene and that of the sensor. The sky background is the dimmest part of a scene, and the brightest level is usually from a bright star or galaxy core. For a sensor's theoretical dynamic range, the well capacity defines the maximum possible signal and the read noise defines the "floor" for the smallest distinguishable signal. (When considering practical dynamic range for a given exposure time and temperature, other sources of noise, like the thermal noise, must also be accounted for in the noise floor.)

Dynamic range is frequently expressed in terms of decibels (dB), a base-10 logarithmic unit. To convert a raw ratio to dB, the formula is

$$DynamicRange(dB) = 20 \times \log_{10} \frac{brightest}{dimmest}$$

Using a logarithmic scale simplifies the math when manipulating values. For instance, if the signal is also measured in decibels, the dynamic range is the maximum capturable signal level minus the noise.

F-stops are another measure of dynamic range, borrowed from photography. This is a base-2 logarithmic scale—an increase of one corresponds to a doubling of dynamic range. For example, a sensor with a read noise of 10 ADU and a full well capacity of 500 ADU has a 50-fold dynamic range. This could also be expressed as 17 dB, or 5.6 stops.

Applying gain reduces the dynamic range captured from a scene because only a portion of the full well capacity is mapped to the ADC's range of digital values. Therefore the brightest value detectable is less than the full capacity. This provides more discrete ADU levels of brightness within this dimmer range at the expense of sacrificing the ability to capture any brighter levels. Any part of the image brighter than gain allows will be clipped, leading to "blown out" highlights in brighter conditions.

Figure 3 and the following simplified example illustrate the point. Arriving photons create the same number of electrons in the well for each of two exposures taken with a DSLR. The well on the left is shown at ISO 100, which for this camera applies no electronic gain, and at ISO 200 on the right, where a 2× gain is applied. With no gain applied, this count of electrons is about 40% of the well capacity, and the 12-bit ADC maps this to a value of 1,600, which will correspond to a dark gray in the image file. When 2× gain is applied, there are now twice as many electrons, which the ADC maps to a value of 3,199, a light gray in the image.

So why bother with electronic gain? Why not just count the number of electrons then multiply the number by two instead? Because it allows us to get a more accurate count when we have few electrons. As an example, let's assume that this hypothetical photosite has a well capacity of exactly 40,950 electrons, so without any electronic gain applied, the full well capacity maps to the ADC's 12-bit range (0–4,095) at ten electrons per ADU. We'll also assume that in each example exposure there are exactly 15,996 electrons generated by photons striking the sensor.

No applying any gain allows us to capture the greatest dynamic range from the scene, but we lose some resolution in our ability to count electrons, since it takes 10 electrons to increase the ADU by one. For this exposure where photons generated 15,996 electrons, the ADC outputs a value of 1,600 (15,996 divided by 10 electrons per ADU, rounded to the nearest integer), but any electron count within +/-5 of 16,000 would also yield 1,600. We just don't have the bit depth to distinguish between them.

If we apply a 2× analog amplification, the 15,996 electrons are approximately doubled by the gain circuit before they enter the ADC. Let's say that the result is 31,991 electrons after a tiny bit of error. The ADC maps this to a digital value of 3,199.

With 2× gain we can only use half the full well capacity, since any more than that will simply map to the maximum value of the ADC—known as clipping. But we can now resolve smaller differences in the electron count, since each incremental change in the ADU now corresponds to five electrons. Finer levels of brightness can be discerned, so intermediate shades are now available that were indistinguishable at a lower gain. We are trading dynamic range for precision. If we know our target will be dim, we can use gain to prevent wasting a lot of our bit depth on unused levels at the brightest end of the dynamic range. However, if there are bright regions of the scene as well, they could be clipped.

The same logic applies to greater levels of gain with smaller and smaller fractions of the well capacity being mapped to the default bit depth through amplification of the charge. You can imagine that at ISO 400 for this camera (4× electronic gain), we already have too many electrons. The ADC will read out at the maximum brightness value of 4,095 no matter how many more photons arrive. We will never know the actual electron count, only that it was greater than a quarter of the well depth. However, if there had been fewer electrons in the well, this level of gain would have mapped the smaller range of photon counts onto the same 12-bits, resolving increments of brightness four times smaller than with no gain at the expense of reducing the dynamic range the sensor can capture by 75%.

In DSLRs, the highest ISO settings (usually ISO 3200 and up) may be different from the others in that the final digital value from the ADC is simply multiplied to create a sort of imitation gain. Because there is no amplification to the analog signal going into the ADC, this reduces dynamic range without providing any additional resolution. It should be clear that the same data manipulation (multiply by two, etc) could be done in post-processing. You can check if this happened by looking at the pixel value in the image—if all the output values are even multiples, then synthetic gain has been applied.

Choosing the best gain setting is a trade-off, but for consumer cameras, there are some special considerations. First, exclude the highest levels on DSLRs if they are not 'real' (usually ISOs of up to 1600 are fine, but it can be higher on premium cameras). Typically the lowest settings are not sensitive enough for astronomical use unless the exposures are extremely long or the target is extremely bright. Also exclude any of the "in-between" levels, like 500 or 1,000, since those can be the result of either two-step amplification or simple multiplication of the ADC's output. For many DSLRs, this narrows the choices to ISO 400, 800, or 1600, depending on the dynamic range of the target and the exposure time. Of course none of this is an issue for dedicated astronomical cameras where the gain is fixed.

The importance of bit depth

For daylight photography, even 8-bits (256 levels) of data for each color channel is more than the human eye can distinguish. If you see "more than 16 million colors" or "24-bit color" advertised for a display, it is referring to 8-bit depth for each of the three color channels. There are 256 levels possible in each of red, green, and blue creates 256 × 256 × 256 = 16,777,216 different color combinations.

Because astronomical targets have a much narrower dynamic range than daylight scenes, we need to capture images with at least 16-bits per color channel, and we typically expand this to 32-bits for intermediate processing steps. Imagine if you took an 8-bit exposure of a faint nebula. Even after several minutes of exposure time, all of the data other than bright stars would be down in the lowest values, maybe the darkest 15 levels. Fifteen discrete shades are not enough to smoothly represent the wispy and subtle nebulosity seen in the best deep sky images. We need to distinguish fine differences in brightness, especially since we are going to "stretch" the image so the nebula's narrow range covers a much wider range of brightness levels. To quantify these differences, we need more resolution between differ-

ent brightness levels. We need more bit depth. The same 15 dimmest levels in 8-bit would be divided into 240 discrete levels in 12-bit data or 3,840 levels with 16 bits, making a much more appealing image with smoother gradients. In short more bit depth in the image data means greater precision, which makes for a better final image.

Even greater precision is required by the calibration process, since many exposures are averaged into one, yielding fractional values that require a greater bit depth to express than any of the original subexposures. In order to retain this level of precision, calculations are done on a floating point basis (i.e. decimal values are used rather than integers), and the final images generated by calibration are typically 16- or 32-bits.

Imagine that after taking a dozen exposures of our hypothetical nebula above, the individual 12-bit values from the camera for one pixel average to 91.52 on the 0–4,095 scale. If an adjacent pixel had an average value of 92.08, they would be indistinguishable in 12-bits, as they both register as 92. A 16-bit file has sixteen times the resolution of a 12-bit file, so it can more accurately express them as the discrete values 1,464 and 1,473 on a scale of 0–65,535.

To illustrate the visual difference that bit depth makes in processing, Figure 4 shows an image stretched with exactly the same series of steps in Photoshop, except the one on the bottom was reduced to 8-bit data before processing, while the one on the top remained in 16-bit. The smooth gradients of the 16-bit image are the result of more data being available to represent smaller discrete changes in brightness, while the larger jumps in the 8-bit image lead to a noisy final image.

Response curves and raw files

The eye's response to light is not linear; instead it is approximately logarithmic. That is, we perceive brightness in steps that get larger as the source gets brighter. Another way to think about it is that we perceive brightness proportionally. We can detect a tiny increase in photon flux if the overall scene is dark, because the change is noticeable in proportion to the overall level of light. The same absolute increase would be imperceptible in a brighter scene where it would be a tiny fraction of the overall flux. This is true for some of our other senses as well—weight, pitch, and loudness are also perceived on an approximately logarithmic basis.

Understanding Images

Figure 4. The same image processed at 16-bits (top) and 8-bits (bottom)

Combined with the variable aperture of the iris, this compression of sensitivity allows us to perceive an enormous range of brightness within a scene, a range that far exceeds that of a printed page or electronic display. We don't think of it this way, but the brightness difference between indoors and a sunlit outdoor scene is on the order of thousands of times. Notably, film's response curve is similar to that of the human eye, while electronic sensors are almost perfectly linear. And just to complicate things, display devices like monitors have nonlinear response curves with a slightly different shape.

When shooting normal photos in JPEG mode on a consumer camera (or looking at the camera's preview screen), the camera tries to accommodate the nonlinear response of your eyes and the display to reproduce the scene as your eye would have perceived it. To do this, it applies a nonlinear transformation (called a gamma curve) to the raw linear data from the sensor. This way, when the final image is shown, the brightness levels on a display device reflect what was originally recorded by the sensor. If you are viewing the image in raw format, this gamma transformation is not applied, so the image on your monitor may not look as bright as it did on the back of your DSLR.

Figure 5. But I thought I took a picture of the Rosette Nebula?

It's hard not to be disappointed the first time you see a raw astronomical image on the screen. The raw linear data are rarely pretty. Perhaps it looks something like Figure 5. Your first thought might be, "What happened to my image?" This is a consequence of the small dynamic range of these images. Look at the histogram of this image, shown in Figure 6. This graph shows brightness levels along the horizontal axis, from completely black to the brightest white. The frequency that each of these levels occurs is shown in the vertical axis. You can see that nearly all of the pixels in this image fall within a very narrow and dim part of the total dynamic range.

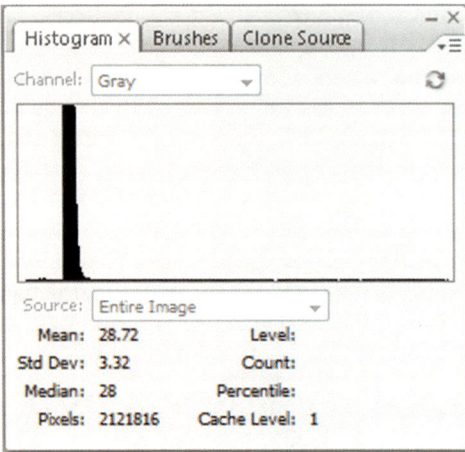

Figure 6. Without processing, the nebula's data are compressed into that tiny spike

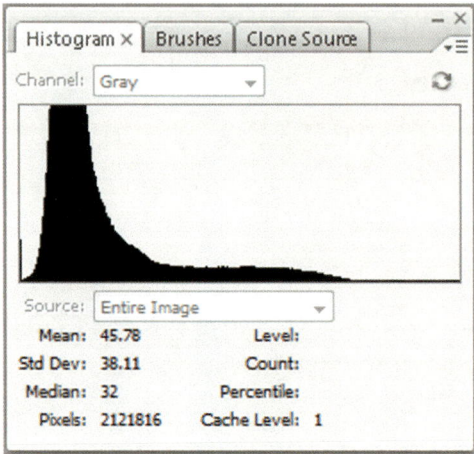

Figure 8. Now the image fills the available dynamic range, revealing the nebula

Fortunately, we have tools at our disposal to bring that dim image up to its full potential. By "stretching" the image in post-processing, we can reveal what's within that spike on the histogram by remapping those few values across a much larger range of brightness. Figure 7 and Figure 8 show the resulting image and histogram. We'll explore this process in detail later.

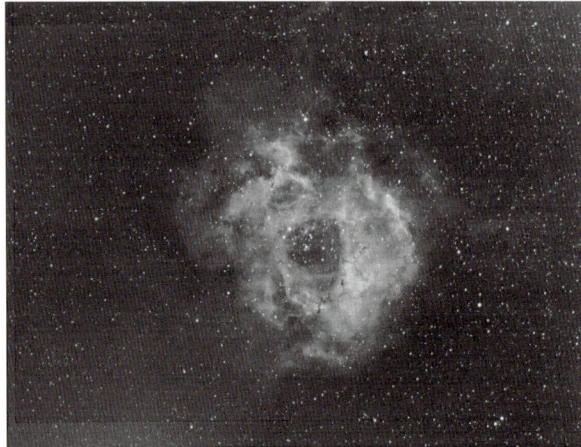

Figure 7. After stretching, the Rosette is revealed

Creating color images

Sensors are inherently color-blind. Since they cannot determine the wavelength of the photons they capture, filters are used to determine color in images. The output of each photosite is a monochrome value representing a level of brightness, but filters allow us to separately quantify how much light is from each portion of the spectrum by allowing only a range of wavelengths to pass through to the sensor. There are two primary methods of accomplishing this: monochromatic sensors with full field filters and "one-shot color" (OSC) sensors with filters built on the individual photosites.

Monochromatic imaging systems have no color filters built onto the sensor. The only limits to efficiency are the inherent qualities of the sensor, but the output is a gray scale image. Colored filters can be placed in front of the sensor to capture the scene in red light, green light, and blue light. These images can then be synthesized as channels into a full-color image at the computer. In addition a luminance channel consisting of exposures taken without a color filter can be incorporated into the image. Monochromatic cameras also allow the user to choose filters other than red, green, and blue. There are narrowband filters that transmit only light from the specific spectral lines that are common to some astronomical objects. As we'll see these have designations like Hydrogen-alpha and Oxygen-III that are based on their atomic source. By blocking all but a small part of the spectrum, these filters can also allow you to capture images through light pollution that would otherwise cause problems. Exposures through narrowband filters can be combined into false color images like those commonly seen in Hubble Space Telescope images. All of this comes at a cost in terms of money and complexity, however. Filters and filter-wheels can be expensive and cumbersome, and there is an additional learning curve involved in properly combining exposures taken through different filters.

One-shot color systems, like DSLRs and some dedicated astronomy cameras, use an array of tiny colored filters over the individual photosites. Each pixel thus determines the brightness of either red, green, or blue light. The true color of each pixel is then estimated using color data from the surrounding pixels.

The most common color array pattern is called a Bayer matrix, and it consists of red, green, and blue filters in an alternating pattern, with 50% green-filtered pixels, 25% red, and 25% blue. This bias is reflective of the fact that green is in the middle of the spectrum, where the human eye has greatest sensitivity.

Figure 9. The Bayer matrix

So how do we calculate the full color information for each pixel when we only measured one part of the spectrum? De-Bayering is the mathematical process of interpolating the true color of each pixel based on the levels in the surrounding pixels. There are actually numerous methods for de-Bayering, each varying in calculation speed and accuracy for a given setting. Smooth gradients are generally easy to de-Bayer, but sharp edges, like stars on a black background, are more of a challenge. The algorithms currently used generally do a good job of creating images that are nearly free of artifacts, so this is something that the typical imager rarely has to worry about. It is important to understand that the color image generated by a color array like the Bayer matrix can only produce an estimate of the true color at each pixel. Typically, the estimate is incredibly accurate, especially for large features, but it is not quite as accurate as a monochrome sensor with full-frame filters. Note that raw files are not de-Bayered, so without proper software to render them, they may look strange in some imaging program. The interpolation of color is usually done by the calibration or post-processing software.

Choosing between one-shot color and monochrome image sensors is about trade-offs between flexibility, cost, and simplicity. Dedicated astronomical cameras, even the one-shot color models, are more expensive than off the shelf consumer DSLRs. The economies of scale simply don't exist for such a niche audience. That said, dedicated astronomy cameras are designed for one purpose, so features like built-in cooling, software control, and low-noise electronics facilitate high quality astronomical images. Monochromatic cameras are incredibly efficient, and some types of work, like photometry or narrowband imaging, can only be accomplished with these cameras. They also lack the infrared filter that is built into most DSLRs, thus they are more sensitive to this region of the spectrum that is crucial for capturing emission nebulae and parts of some galaxies. The well-depth of dedicated cameras is typically much deeper as well, allowing longer exposures and greater dynamic range.

DSLRs, despite their limitations, are cheaper, use larger image sensors than equivalently-priced astronomical cameras, are generally easier to use, and are supported by a range of accessories. Their built-in preview displays facilitate quick focusing and composition. Noise levels are low, and they are improving with every new model. There are no filter wheels to add, and no additional processing steps to combine color channels. Being consumer devices, they have compromises, but the economies of scale in production make them a great value. For color imaging on a budget, they are hard to beat, and the results some amateur imagers have gotten with them speak for themselves.

Sensor architecture

There are many different technologies available for electronic imaging, so it's good to know a little about the strengths and compromises involved in sensor design. First, we've seen that there are two basic types of imaging sensor technology: CCD (Charge-Coupled Device) and CMOS (Complementary Metal Oxide Semiconductor). While the details of the two technologies differ, the end results are very similar. CMOS photosites have active electronics on each photosite that reduce the area available to collect photons, but allows each photosite to be read individually. They require far less power to operate than CCDs, and they are also cheaper to manufacture. The CCD readout process requires the electrons from each photosite to be passed between rows or columns like water in a bucket brigade as they are quantified, but CCDs are generally more sensitive, with special back-illuminated CCDs boasting 90% or higher quantum efficiencies.

Within the CCD family, there are also different shutter technologies. Full-frame CCDs use the whole surface to gather light, so they need a mechanical shutter to block the sensor while it is read. Without a shutter, photons continue to strike the sensor while it is being read out, smearing the image. Frame-transfer and interline sensors mask part of each photosite or alternating rows, respectively, as an interim storage place for electrons while the sensor is read, so there is no need for a mechanical shutter. Some of the sensor's area is lost with these technologies, so there is a cost in resolution and sensitivity, though microlenses can help direct light away from the masked areas.

Another parameter to consider with astronomical CCDs is the presence or absence of anti-blooming gates. These prevent an overflow of charge in one photosite from bleeding over into others during the readout process. This manifests on the image as vertical streaks or smears from the brightest stars, causing irretrievable loss of data in the surrounding pixels. Anti-blooming gates allow overflow electrons to collect without spilling into adjacent photosites. Since they mask part of the photosite area, these gates reduce the sensitivity of the sensor. Though this sensitivity problem has been partially overcome with advances in the use of microlenses, CCDs without gates still have greater sensitivity. Users of non-anti-blooming gate CCDs must be careful not to let individual exposures go long enough to saturate any photosites. This is a problem specific to CCDs; CMOS sensors do not suffer from blooming.

2 SIGNAL AND NOISE

Signal and noise at the pixel level

Signal and noise are possibly the most important concepts to understand about astronomical imaging, and nearly every step of image capture and processing is done with them in mind. In imaging, the signal is anything that causes the output of the sensor to increase in a time-dependent manner. This is mostly from photons, but heat can be an additional, unwanted source of signal. While noise isn't something that most daylight photographers think about much, that's because they don't spend hours teasing out faint wisps of signal from a nearly dark image. And therein lies a crucial insight: when things are dim, noise is more of a problem than when there is plenty of signal.

But what exactly is noise? You know a noisy image when you see one, but what is it and how does it relate to signal? Noise is uncertainty—it makes an image's brightness value fluctuate randomly around its true value each time we measure it. Noise is the amount of random variation that a signal exhibits, and the less noise there is, the better.

In film photography, noise manifested as graininess. In digital imaging, a noisy look stems from variation where there shouldn't be any, where regions of solid color or brightness showing random pixels at levels above and below the expected value.

It is important to understand that noise is not a distinct portion of the signal nor is it an impurity that we can separate from the signal. It is a measure that describes the accuracy of our measurement of the signal. It's analogous to the margin of error in a poll, and just as larger polls have smaller margins of error, longer total exposure times give us a better estimate of the signal.

It is useful to describe the amount of noise in a signal as a proportion of the signal itself, but bear in mind that the noise is not a separate entity. The most commonly used expression for this measure is the signal-to-noise ratio, abbreviated SNR or S/N, which is possibly the most important measure of image quality. Images with a higher SNR have less visual noise, smoother gradients, and they can be stretched further in post-processing to reveal fainter and more subtle details. The same quantity can also be expressed as N/S, known as the coefficient of variation, which is simply the amount of noise as a proportion of the signal.

Signal and noise are inherently mathematical concepts, so some math is needed to explain them, but we will not delve into the detailed mathematics any more than necessary. It is only important to understand the big picture.

The underlying goal of deep-sky imaging is to accurately capture a count of photons coming from an incredibly dim object, probably against a light-polluted background. We achieve accuracy through long total exposure times and the use of what are called calibration frames. Long integration times, usually divided across many exposures, allow us to increase the ratio of signal to noise, while calibration frames allow us to correct for effects on the image that come from the sensor, optics, or unwanted signals.

There are two sources of noise in images:

- Noise that comes from the random fluctuations of a signal. We call this **shot noise**.
- Noise that comes from the electronic processes used to measure light. We call this **read noise**.

Of the two, shot noise is almost always the larger problem, and we'll cover it first.

Even though we only seek to capture photons from our astronomical object, there are actually three signals detected by the sensor, as shown in Figure 10. First, there are the photons from a distant astronomical object that we are interested in. Second, there are the unwanted photons from the skyglow in our atmosphere. Third, the ambient heat of the sensor contributes an unwanted thermal signal to the image, dislodging electrons into the sensor wells without the presence of photons. For each of these signals there is some variation in the exact number of photons that arrive with each exposure, thus each is a source of shot noise.

In addition to the noise in the three signals above, there are two sources of noise that are independent of any signal. They come from the electronic processes of the camera. First, there is some uncertainty inherent in counting the electrons in each photosite, which is called read noise. Secondly, there may be some rounding necessary when this value is converted to a digital value from the ADC's finite bit depth. This is called quantization noise or quantization error. It is usually insignificant in comparison to the other sources of noise, and it can be considered part of the read noise.

Additionally, each photosite is not identically sensitive to light, which leads to two-dimensional variation that we have to correct with calibration frames, which we'll review later.

Figure 10. Three signals are captured in each exposure

Shot Noise

Let's start with a hypothetical camera that has no read noise—that is, it will perfectly count photons and transfer an appropriate digital value to your computer without adding anything extra. Further, this camera is sitting out in space with a perfectly dark sky and almost nothing between it at the object you'd like to image. It tracks perfectly, never deviating from where you point it, capturing ancient photons coming from exactly the same region of space. In this thought experiment, you have your own space telescope with a perfect camera. Lucky you!

Even in these ideal conditions, however, there is still noise to deal with because light is not a continuous quantity—it is quantized into discrete photons. No matter how steady the source, these photons do not arrive at a perfectly constant rate. Raindrops are a similar phenomenon. A steady rain falling at a rate of one centimeter per hour does not reach that average rate with the raindrops falling in lockstep at precise intervals. Knowing that the long-term rate is a centimeter per hour does not tell you how many drops will splash into your rain gauge over the next ten seconds.

To continue the raindrop analogy, imagine your sensor as a floor of square kitchen tiles, each tile representing a photosite. Photons rain down at a steady rate, striking randomly all over this sensor floor. After a set time, you close the shutter and go out to count how many have hit a given tile. Let's say you count 33. You then get out your photon squeegee to wipe the sensor clean and open the shutter again for the same amount of time. Do you think the count will be 33 again? Not likely. It will probably be a nearby value like 28 or 40 or 34. This is the noise inherent in the rate of photon arrival. We call this fluctuation "shot noise," and it is a consequence of the fact that light energy arrives in discrete particles. (The term comes from the analogy that photons are arriving like the scatter of shot pellets from a shotgun.)

When you photograph objects during the daytime, this is rarely an issue. If in one second a flood of 100,000 photons strike a photosite, but in the next second it's 100,344, the difference doesn't matter much in the final image. But for exceedingly dim objects like those in the night sky, where we might be expecting on the order of a hundred per minute, it starts to matter a lot.

Photon arrival is a "Poisson process," which means that individual photons arrive continuously, and the timing of each arrival is independent of the others. This means that we can predict that the number of photons captured in a set of equal exposures will follow a Poisson distribution. This is fortunate because the Poisson distribution has some nice characteristics that make the mathematics relatively simple. Figure 11 shows a probability distribution where the height of each bar represents the probability of getting that value from a Poisson process whose true mean is 100.

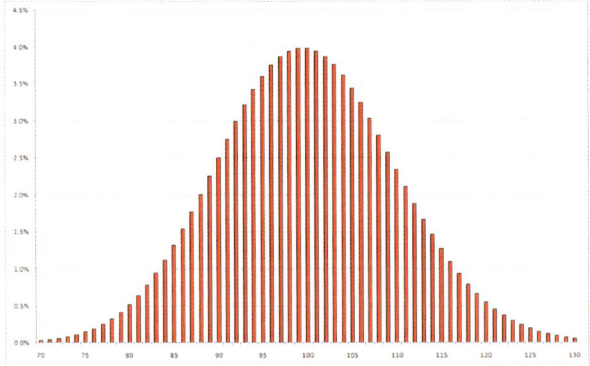

Figure 11. The Poisson distribution for a mean of 100

The standard deviation (SD) is the measure of the uncertainty in our estimate. Graphically, it is a measure of how wide the bell-shaped distribution is. Mathematically, it tells us how much variation there is around the long-term average rate. For a Poisson distribution, it is calculated very simply: the SD is simply the square root of the rate. Since the standard deviation is our yardstick for uncertainty, this is how we quantify noise. The larger the value, the wider the spread around a mean, which indicates more noise.

So what does a standard deviation really mean? The Poisson distribution has a specific shape (very similar to the "normal" or Gaussian curve that underlies most statistics), and it allows us to predict that about 68% of the time, the value of a sample will fall within +/- 1 SD of the true rate. About 95% of the time, it will fall within +/- 2 SDs, and more than 99% of samples should be within +/- 3 SDs.

When we think of the brightness of a point in the sky, we are thinking of the average rate for photon arrival for that spot. Barring an event like a supernova, this rate should be steady over time. By taking multiple exposures of that point, we get many samples with which we can estimate the long-term rate.

Consider a situation where the real long-term average rate is 100 photons arriving per minute. Here, the shot noise is 1 SD = √100 = 10. Based on that, you can expect that about 68% of the time you will count between 90 and 110 photons per minute, 95% of the time the count will be between 80 and 120, and only very rarely will there be less than 70 or greater than 130. The values will cluster around the true rate of 100.

For comparison Figure 12 shows the Poisson distribution for a mean of 1,000. As in Figure 11, the horizontal scale shows 30% below and 30% above the mean, but you can see the distribution is narrower. Here the shot noise is 1 SD = √1000 = 31.6. While the actual width of the bell curve is nominally greater than our example where the mean was 100, as a percentage of the mean, it is narrower.

This is a critical point. When the mean was 100, the signal-to-noise ratio was 100/10 = 10. Now that the mean is 1000, the SNR is 1000/31.6 = 31.6. The noise grows at a slower rate than the signal. We improve our SNR as either a) the object is brighter, or b) we capture more photons through longer total exposure time.

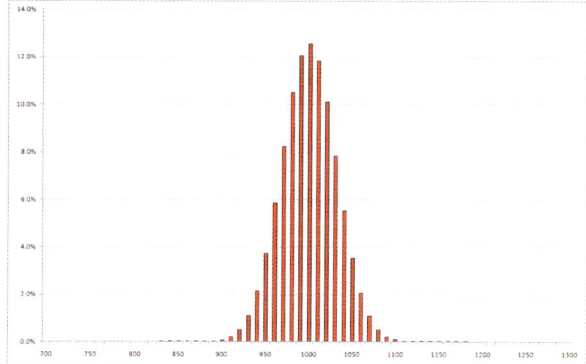

Figure 12. Poisson distribution with mean of 1000

Another way to visualize the concept of noise is to line up 100 random samples and see how they vary. Figure 13 shows this for a mean of 100 and a standard deviation of 10, just as above. The range of +/- one standard deviation is highlighted in blue. You can see that most of the values stay within one SD of the mean, but there are occasional outliers. Overall, it's a noisy picture, and with only a few samples, it would be hard to guess that the mean was 100.

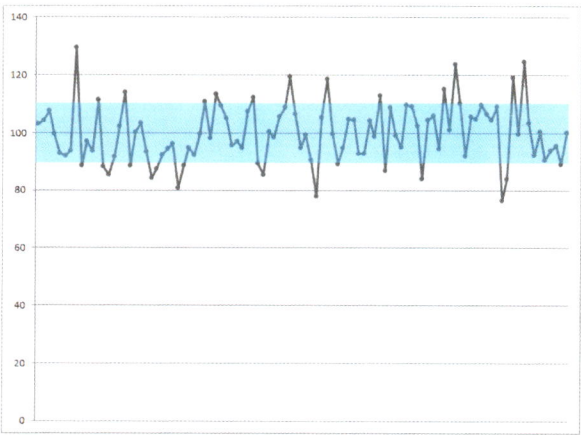

Figure 13. Shot noise with a mean of 100 (+/- one standard deviation shown in blue)

Compare this to Figure 14, where the mean is 10,000. The actual shot noise is 10 times greater here (√10,000 = 100), but it looks smoother because noise is a smaller proportion of the total. The SNR is now 100, or stated another way, the noise is now 1% of the total.

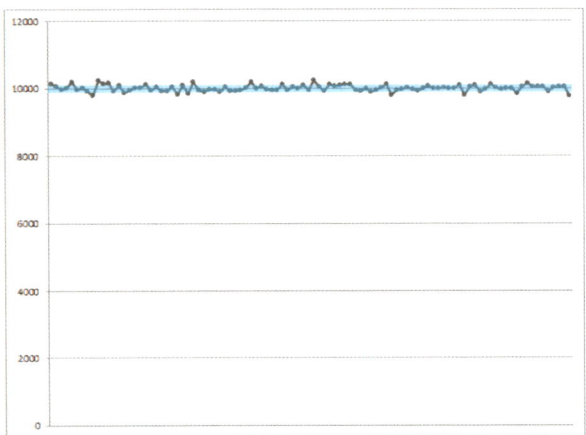

Figure 14. Shot noise example with a mean of 10,000 (+/- one standard deviation shown in blue)

> Credit for discovering the Poisson distribution goes to Siméon-Denis Poisson, in his 1838 book *Research on the Probability of Judgments in Criminal and Civil Matters*. Astrophotography was not exactly what he had in mind, but his results have since had enormous impact across many scientific disciplines.
>
> The normal distribution is an excellent approximation of the Poisson distribution for large values, so for convenience, we assume that they are equal here when calculating probabilities.

An example of shot noise

Let's look at another illustrative example. We'll again zoom in on a single pixel from one photosite on the sensor, using our nearly perfect camera. We'll ignore all the others and consider this one lonely pixel gathering light from a small patch of the sky.

Imagine that for this one pixel, the true long-term average arrival rate of photons is 1,000 photons per minute. Because of shot noise, exactly 1,000 photons don't arrive every minute. Sometimes it's more, sometimes it's less, but the average rate is 1,000. For now, we'll ignore how the camera counts the photons, assuming again that our camera is perfectly accurately.

If we only take one exposure, that will be our best estimate of the true value. Let's say we take a single one-minute exposure, and upon counting the photons, we find that 985 have arrived. Remember, we don't actually know the true value is 1,000. For all we know, it is 1,050 or 962 or 983. At this point, our best guess is that it's 985.

Since we know that photons don't arrive regularly, we'd like to be more confident in our estimate. So we take another exposure, and of course we get a different count this time: 1,059.

This means that over two minutes, we've counted a total of 2,044 photons arriving from the same point in the sky, for an average rate of 1,022 photons arriving per minute.

Individual Exposures	Total Photon Count	Average per minute
985	2044	1022
1059	(1 SD uncertainty = +/- 45.2)	(1 SD uncertainty = +/- 22.6)

While our estimate has changed, what's really important to notice here is the level of uncertainty *as a proportion of the total value*. The first exposure had a shot noise of $\sqrt{985}=31.4$. This was about 3.2% of the total, or alternately a SNR of 31. (As you have probably noticed, when we only have shot noise to deal with, the SNR is algebraically equal to the noise level. The situation will get more complex when we add other sources of noise.)

When we add the two exposures together, we get a total count of 2,044 photons over two minutes of exposure time. The square root of 2,044 is about 45.2, so that is our 1 SD uncertainty around that value. We divide by two to get a per-minute rate of 1,022 photons, which is our new estimate of the true value. We also divide the uncertainty by two, for a new shot noise level of 22.6. As a proportion of the total signal, the noise is now only about 2.2%, or an SNR of about 45. Since shot noise grows with the square root of the signal, combining two exposures reduces the noise by the square root of the number of exposures; here, $\sqrt{2} \approx 1.4$ fold.

Capturing more photons improves the signal-to-noise ratio, whether this is done through a combination of multiple exposures or one long exposure. With our perfect camera, we could have accomplished the same result with a single two-minute exposure if we chose.

This makes sense when you consider that as the signal grows linearly, the shot noise only grows by the square root. Since the noise grows more slowly, it is always falling further behind. Considering only shot noise, the signal-to-noise ratio grows with the square root of the total exposure. In other words, to cut your noise in half (double the SNR), you'll need four times the total exposure. To reduce it by 75%, you'll need 16 times as many.

Understanding Images

If two one-minute exposures are good, four must be even better, so we go out to our telescope and take two more. If the math holds, we should expect to get a $\sqrt{4} = 2$-fold reduction in noise as a proportion of signal compared with a single exposure. This should double our signal-to-noise ratio.

Individual Exposures	Total Photon Count	Average per minute
985		
1059	4037	1009.25
991	(1 SD uncertainty = +/- 63.5)	(1 SD uncertainty = +/- 15.9)
1002		

And indeed, our average value is starting to converge toward what we know is the true value as our uncertainty gets smaller. The SNR is now up to about 1009/15.9 = 63.

Based on these four samples, we revise our best estimate of the true value to 1,009.25, which is the average of our four values. At this point, we can be reasonably confident that the truth lies somewhere close to 1,009.25.

The night is still young, and we know we can sleep in tomorrow, so we decide to double our exposure count to eight. (There is nothing special about the exact number of exposures; powers of two were chosen here just to make the math a little clearer.) It should be apparent what sort of result to expect by now.

Individual Exposures	Total Photon Count	Average per minute
985		
1059		
991		
1002	7972	996.5
958	(1 SD uncertainty = +/- 89.3)	(1 SD uncertainty = +/- 11.2)
1020		
984		
973		

Now our average is 996.5. As we add more exposure time, we can be confident that our estimate of the rate of photon arrival gets more accurate. This reduction in uncertainty is the same as saying that we are reducing noise as a proportion of the signal. Our SNR for eight minutes of exposure time is now 89.

The math is all well and good, but what does this reduction in noise mean in the final image? Imagine that our example camera above were pointed at a nebula. Instead of the single pixel we've been considering, let's also look at the three next to it as well. The tiny area of the nebula these four photosites are aimed at has a very subtle gradient where it is slowly getting brighter. Our example photosite had a true flux of 1,000 photons per second, and the three next to it have true fluxes of 1,001, 1,002, and 1,003. In a final image, the ability to accurately show subtle differences in brightness makes the image look sharp and realistic. In digital imaging a noisy look stems from variation where there shouldn't be any—pixels with brightness levels above and below the expected value. By improving the signal-to-noise ratio, we can capture more subtle differences in brightness which can reveal structure at finer scales. In short, noise sets a limit on the smallest change in brightness that can be captured in your image.

The table illustrates the visual effect of noise in stacks of images of the galaxy M101, from a single four-minute exposure to the average of 64 such exposures. (Again, any number of exposures could be taken; powers of two are only used for clarity.)

The same basic stretch was applied to all of the images, no other processing was done. The images on the left show a view of the whole galaxy, while those on the right are a tighter crop to show the noise in detail. The region of the crop is the same in all exposures, bounded by the green square.

The noise overwhelms everything but the largest features of the galaxy in the single exposure, since only those macro features have enough difference in brightness and spatial distance to show up. Spatial distance also plays a role in images since the brain can effectively average the brightness of whole regions of pixels, which is a way of sacrificing spatial resolution to improve SNR. Imagers achieve the same effect through the "binning" of multiple photosites to act like one. A 2×2 group of pixels will receive four times the number of photons, and thus have half the shot noise of a single pixel, but at the expense of resolution. The brain does something similar when you look at an image.

By the time we reach 64 exposures, subtle differences in brightness become apparent in the spiral arms, even in this minimally processed image. Finer spatial details become clear as well. The noise is about eight-fold lower than in the single frame, but you can see that diminishing returns are setting in. To cut the noise in half again would require almost two hundred more exposures. Since these were four minute exposures, that would move this into a multi-night project.

By now it should be clear why reducing noise is our goal. Higher SNR means better estimates of the true value at each pixel, which means smoother gradients and more detail in the final image.

Before moving on to other sources of noise, let's reconsider the effects of reduced signal with a second example. In this one, the camera is getting a lot less photons. Maybe this is a dimmer part of the object. Or maybe we are taking shorter exposures. Or perhaps we had to sell our old scope and downgrade (shudder!) to slower optics. In any case, the long term average brightness is now 100 photons per minute striking the photosite instead of 1,000.

As before, we have no idea what the real value is, so we take an exposure to try to find out. Our first shot shows 108 photons. Being omniscient in this experiment, we know that the true brightness should be 100, so why is this first shot 8% off? Was this an outlier event? Not at all. For a Poisson process, the standard deviation is the square root of the value, and the square root here is 10.

So despite being so far from the true rate, 108 was a reasonable value to expect. Proportionally, the noise component of this dim signal is 10% of its value, whereas our brighter example had a noise level of only 3% for a single shot. When the signal is small, noise can be uncomfortably high in relation to the overall signal.

Just like before, we decide to take another shot and average the two.

Understanding Images

Individual Exposures	Total Photon Count	Average per minute
108	209	104.5
101	(1 SD uncertainty = +/- 14.5)	(1 SD uncertainty = +/- 7.2)

Two minutes of exposure increases our SNR from about 10 to 14. As before, let's double our exposure count again.

Individual Exposures	Total Photon Count	Average per minute
108		
101	415	103.75
118	(1 SD uncertainty = +/- 20.4)	(1 SD uncertainty = +/- 5.1)
88		

And as before, taking four shots doubles ($\sqrt{4}= 2$) the SNR we got from one exposure, but this combined SNR of 20 is still worse than even a single exposure of the brighter object, where the SNR was about 31.

Noise is larger in relation to the total in dim exposures. It would take far more exposure time to get the same level accuracy as with our brighter example. In fact, it would take 10 times as many samples at one-tenth the brightness, because with our perfect camera, all that matters is the total number of photons we capture.

To overcome shot noise, we need to collect more photons. We have several options to do this.

- We can gather more light in each exposure. This means we either take longer exposures or we increase the aperture of our optics while holding the focal length constant (reduce the focal ratio).
- We can take more exposures.
- We can capture a larger area of the sky with each pixel (known as our image scale). To do this we either use a sensor with larger pixels or we use optics with a shorter focal length.

Skyglow

Now let's take a step toward the real world. We still have a perfect camera, but our skies are more realistic. We've come back from orbit, and the sky is not perfectly dark any more. We have another signal coming in: a glowing (probably orange-brown) radiance has replaced our beautiful pitch-black space. Now the photons from our target aren't the only signal: we have skyglow to deal with.

We really had a good thing going there in outer space. Even in the darkest spots on earth, far from human-made light pollution, there is still a small amount of atmospheric glow from ionization in the upper atmosphere, not to mention dust particles, humidity, and the natural light scattering properties of air. Though it varies tremendously between locations, skyglow is unavoidable.

Since we have powerful image processing tools at our disposal, and our first thought might be, "we can just subtract that skyglow out!" It should be pretty consistent across the field of view, so we'll measure an empty spot in the image, and take that value away from every pixel, leaving a nice black background. It's genius, a plan that will restore our outer space view, and prevent our images from looking like nonpareil stars on a chocolate sky.

All true, but not so fast. What about noise? The skyglow is its own separate signal, thus it has its own noise. This means even if we subtract the skyglow, its noise remains. Noise is an inherent quality of the signal, not a separate entity, so it can't be subtracted. To make matters worse, in some cases, the skyglow is brighter than our object. Let's have a look at this effect.

We now know that we need a lot of photons to get a decent estimate of a pixel's brightness, so we take four minutes of exposure to start with. Our hypothetical dim target is still emitting an average of 100 photons per minute, and our camera is still perfectly counting each one. But now we have skyglow, and it is five times brighter than the object. (Apparently, our return from orbit landed us in suburbia.) With each exposure, we are getting around 600 photons: 500 from the skyglow and 100 from the target.

Here are our four exposures with skyglow:

Individual Exposures	Total Photon Count	Average per minute
573		
614	2424	606
630	(1 SD uncertainty = +/- 49.2)	(1 SD uncertainty = +/- 12.3)
607		

We know that photons are striking the sensor both from our target object millions of light years away and those from light scattered in the atmosphere, but it has no way to tell them apart. They are all just photons, and there is no way to distinguish them.

Based on four exposures, we estimate the combined flux of the target and the skyglow to be 606 photons per minute.

And we know that the noise component in this is about 12.3 ($\sqrt{2{,}424}/4$), for an SNR of 49, which seems pretty good.

But we're savvy with our image processing, so we subtract 500 from the photon count in each pixel to correct for the skyglow. While this brings the background back to the level it was in our picture from space, the image looks much noisier. The problem is that while we can subtract the signal from the skyglow, we cannot subtract the noise that came with it. No matter what we do to the final total mathematically, the uncertainty in the estimate remains.

The noise from collecting a total of 2,424 photons is about $\sqrt{2{,}424} \approx 49.2$. That looks fine, because it's only 2% of the total. But the only signal that we are interested in is that from the target, and those were only about 400 of the total (we don't know exactly how many). A noise level of 49.2 is about 12% of this signal. That's an SNR of 8, which looks far from fine. The larger combined signal of the object and skyglow fluctuates over a larger range than the signal from the object alone, and this fluctuation remains even when we subtract from the total. The weaker our target's signal is in relation to the skyglow, the worse the effect. We have to shoot longer and longer total exposures to overcome this difference as either the skyglow gets brighter or the targets get dimmer.

Now let's briefly consider the same example, but at a dark sky site, with 10% of the skyglow as before, a mere 50 photons per minute of flux through our optical system. As before, we take four exposures.

Individual Exposures	Total Photon Count	Average per minute
137		
155	591	147.8
146	(1 SD uncertainty = +/- 24.3)	(1 SD uncertainty = +/- 6.1)
153		

Because the object's flux is a larger proportion of the total, the skyglow's noise doesn't swamp it. The SNR for the total signal is 24.3, which seems lower than the 49.2 we got at our urban site. But when we consider it relative to the signal we are interested in, it compares much better: 400/24.3 = 16.5 versus the 8 we saw with brighter skyglow.

No matter where we are imaging, there is noise coming from non-target sources, including skyglow. The only way to overcome that noise is longer total exposure time. With this reduced noise, we allow the object's light flux to be distinguishable from the skyglow's flux in post-processing.

To be clear, skyglow is the dominant source of noise for urban and suburban imagers. If you are under sixth magnitude skies or capturing narrowband images, other sources of noise can become dominant, but skyglow is the primary bugaboo for most imagers. Light pollution filters can help some, as can choosing targets away from the worst pollution. (And join the International Dark-Sky Association if you want to help combat the problem.) Narrow-band imaging is an option that avoids most light pollution by filtering out all but the specific wavelengths emitted by some objects. It works exceptionally well for emission nebulae, but helps little for broadband targets, and it requires long exposures through a monochrome camera with specific filters. For the most part, brighter skies mean that exponentially more exposures are required to reach the same signal-to-noise ratios you could achieve under dark skies. Imaging at a dark location is one of the fastest ways to improve your image quality.

Almost as bad as its impact on SNR is the effect skyglow has on dynamic range. While the theoretical dynamic range of a sensor is fixed for a given gain setting, the *usable* dynamic range that is left for target object photons is reduced with every skyglow photon in the well. If the well is full of skyglow-generated electrons before we've captured enough data in the dimmest areas of our object, the project is hopeless.

Thermal signal, hot pixels, and dark frames

There is a third signal we collect while distant photons hit the sensor, but this one is coming from a much closer source. The warmth of the air and the heat generated by the camera's own electronics cause a small amount of charge to be generated as electrons are agitated out of the silicon substrate. This thermal signal is sometimes referred to as the "dark current." There is no way to tell whether an electron in the well was dislocated by an extragalactic photon or the heat energy in the sensor itself; the ADC counts them all the same. Without some form of cooling, this heat can become a significant source of signal (and thus noise) on all but the coldest winter nights. Like any other signal, it is linearly time-dependent: longer exposures not only capture more photons from that wispy nebula, they are also generating more electrons from the ambient warmth. Dark current is also temperature-dependent. For most sensors the thermal signal doubles with approximately every 6–7° C (11–13° F) increase in temperature. Just like any other signal, it doesn't arrive perfectly regularly, so it adds its own shot noise to the image. Worse, there is variability between photosites in the sensitivity to temperature.

Reducing thermal signal, and thus any associated noise, is the goal of thermoelectric cooling systems. Cooling the

Understanding Images

sensor 30°C below the ambient temperature reduces the thermal signal by about 97%. Deeper cooling can reduce it even more. This makes an enormous difference. Consider the popular Kodak KAF-8300 sensor as an example. On a warm summer evening at 25° C, the sensor's photosites will each produce around 2.5 electrons per second. For a five-minute exposure, that would be about 750 electrons. Each photosite has a well capacity of about 25,000 electrons, so almost 3% of that capacity would be filled with thermal signal alone. Cooling the chip by 30°C drops the thermal signal to about 25 electrons.

The predictability of thermal signal makes its effects a little easier to manage than other sources. By capturing a set of images, called dark frames, that collect only the thermal signal, we can estimate the total dark signal for each pixel and subtract it from the final image. To get an accurate estimate of its value, we must take many dark frames and average them. Dedicated astronomical imaging systems with active cooling may not require dark frame subtraction at all, but uncooled cameras, especially DSLRs, require them at most temperatures.

Dark frames must match the duration and temperature of the exposures used to capture the object (the "light frames"). Without temperature-controlled cooling, the most basic method is to take dark frames in the same imaging session as the light frames, alternating taking lights and darks. While this yields darks that accurately match the temperature of the lights, it is not a recommended approach since cuts the imaging time in half. Clear sky time is precious, so a better option is to create a library of dark frames taken at known temperatures, gains, and durations. If you consistently use the same gain and exposure times, it is simple to make good use of a cloudy night to take dozens of dark frames at one temperature that you can average to create a master dark frame for a given temperature. A large collection of dark frames taken at different temperature will match with exposures from many nights without wasting precious clear sky time. It is crucial to be sure that the temperature of the *sensor* is matched, since the ambient air temperature is not always well-correlated with that of the inside of the camera. Later, we'll detail the process for acquiring reliable dark frames.

The third option is to utilize software that can scale a master dark frame (an average of many darks) to match any exposure time or temperature. Dark current is a time- and temperature-dependent signal, so a master dark frame can be mathematically scaled to match a higher or lower (preferably, lower, since it's more accurate to scale from a higher value to a lower one) temperature or exposure time as long as you know the bias signal of each pixel. The bias is the amount of signal that a pixel outputs when it hasn't captured a single photon. It's the zero point for the pixel, and we'll discuss it in the next section on read noise.

Fortunately, most of the pixels in modern sensors behave well at normal temperatures, so thermal signal is usually manageable. Unfortunately, a fraction of the pixels are "hotter" or "colder" than the others in that they are more or less sensitive to thermal signal. These hot pixels and cold pixels can be orders of magnitude more sensitive to heat, and they look like speckles in an image, especially after it is stretched in normal post-processing. Dark frames correct for this two-dimensional noise component as well.

Figure 15. Comparison of hot pixels at 70° F and 55° F (13° and 21° C)

Figure 15 shows a close up from the same area of two four-minute dark frames from an uncooled DSLR. The one on the left was taken at 70° F (21° C). The one on the right was taken at 55° F (13° C) using the same came at the same gain of ISO 1600 and exposure time of four minutes. Both frames were equally stretched to make the hot pixels more visible.

Most of the effect of thermal signal is evident in the hot pixels (gray or white here, with RGB values of 50–200 on an 8-bit scale), while the majority of the pixels are black in this image (with values around 1 or 2). Also note how much less noise there is at the cooler temperature. The hot pixels essentially disappear for this camera (a Canon Rebel XSi/450D) at temperatures below 30° F (-1° C).

Since hot pixels are the primary source of thermal noise, at colder temperatures (near or below freezing), many imagers with uncooled cameras have had success using a calibration technique called hot pixel mapping in lieu of dark frames. This process algebraically replaces the values of hot pixels based on the surrounding pixels. Hot pixel mapping is available in most calibration software.

> Many DSLRs have a long-exposure mode that automatically takes and subtracts a dark frame from the picture. So why not use this convenient feature? First, only subtracting a single dark frame leaves a lot of noise behind. Thermal signal is very small, therefore noise is a large proportion of it. To effectively reduce the amount of noise added to the final image usually requires averaging dozens or even hundreds of dark frames.
>
> More importantly, taking dark frames is wasted time when you could be collecting more light from the target. More target signal is always better. Even without any dark frames, your final image will have less noise if you spend the entire night collecting light frames (exposures of the sky) than if you spent half the night with the shutter closed for dark frames. Better yet, cloudy nights are perfect for collecting dark frames, since matching the temperature and exposure time are the only parameters that matter there.

Read noise and quantization noise

Unlike our hypothetical camera from the examples, a real sensor's photosites are not perfectly accurate in their ability to count photons. There is some uncertainty introduced not only by the physics of the sensor itself, but in the process of reading the sensor. There are small voltage fluctuations inherent in all electronics that introduce a tiny bit of noise, which is called read noise.

Since it is a fixed level of noise that occurs each time the sensor is read out by the ADC, read noise imposes a sort of per-exposure penalty. The amount of read noise produced by modern cameras is fairly low, and it is primarily dependent on the amount of time allowed to read out the sensor, a value that is generally hard-wired into most cameras. Typical read noise levels are on the order of 5–20 electrons per exposure.

Quantization noise is a fancy name for a simple concept—the ADC produces integer values, so in nearly all cases, some rounding is required when translating electron counts from the well into an ADU value that can be stored in an image file. When the ADC's useful bit depth is less than the full well capacity, there aren't enough digital values to express all possible electron counts. Some voltage levels will have to be lumped together within the same ADU value because it takes more than one electron to yield an increase of one in the final digital value.

If the useful bit depth of the ADC is greater than the full well capacity is set lower, the quantization error is reduced, though there is always some rounding unless the bit depth is exactly equal to the well capacity.

As an example, imagine a sensor with a full well capacity of 10,000 electrons exposed at a level of gain that uses this whole capacity. The analog voltage output from the sensor could be any number of electrons between 0 and 10,000. If this camera has a 12-bit analog-to-digital converter, there are only 4,096 different values of output, so it will take about 2.5 electrons to increase the ADU value by one. Because we can't distinguish between 8,500 electrons and 8,501 electrons, that is a form of uncertainty, and uncertainty is noise. For most practical purposes, quantization error is typically too small to worry about as a separate source of noise since it is typically dwarfed by other sources, but it points to the need to use an appropriate amount of gain.

Both read noise and quantization noise are uncertainties introduced by the electronic image capture process. Unlike shot noise, they are independent of a signal. With reasonable exposure times and proper calibration, these small sources of noise can almost be rendered negligible.

Modern cameras have vastly reduced read noise, and even better, read noise can be partially corrected with bias frames. Bias frames are essentially zero-photon exposures, and they are used in the calibration process to determine the zero point for each pixel. To collect bias frames, the camera is shuttered or covered completely, and the shortest exposure possible is taken with the same gain as the light frames. The idea is to take an exposure that represents the output of the sensor before any photons hit it.

Why does the zero point vary? Most cameras add a small, constant amount of current to that from the sensor well as it goes into the ADC. Since read noise can cause variations that add or subtract from the true charge, this "offset" or "bias" signal prevents a nearly empty well from reading out with a negative charge due to read noise. All such small value would be counted as zero ADUs, making it impossible to distinguish between them. The level of the bias signal is very small, typically less than 1% of the dynamic range.

Ideally, variations in read noise will yield a completely random pattern across the image in each exposure. This way, we can average them out in calibration to produce a smooth final output. Sometimes, however, the electronics of the camera produce fixed patterns, usually horizontal or vertical lines, in the noise that are an artifact of how the photosites are read. This can generally be corrected in calibration, since bias and dark frames will also have the same pattern. In the worst cases, the camera's electronics can inject variable patterns that move from exposure to exposure. This is not usually the case, but when it is, only large numbers of exposures can properly average it out.

> **Exercises**
>
> 1.1 If a sensor's dark signal has a 7° C doubling temperature, and the ambient temperature is 20° C, by what percentage is the dark signal reduced when the sensor is cooled to -20° C? By what percentage is the dark noise reduced?
>
> 1.2 The KAF-8300 sensor specifications state that it has a thermal signal of no more than 200 electrons per second at 60° C, a doubling temperature of 5.8° C, and a full well capacity of 25,500 electrons. What is the thermal signal at 20° C? At that temperature how long would it take for the well to fill with thermal signal? (Assume a bias signal of zero.)
>
> 1.3 Assuming this same sensor has a read noise of 10 electrons, what is the available dynamic range for a 20-minute exposure at 20° C? And at -15° C?

Two-dimensional corrections

In addition to noise at the pixel level, there are two types of two-dimensional defects that need to be corrected across an image: non-uniform sensitivity between pixels and uneven illumination that results from vignetting or obstructions in the optical path. These are corrected with the final type of calibration frame, flat frames. Flats correct for variation in brightness across the entire field of view by showing what an evenly illuminated view looks like through the same optical train used for the lights. The calibration software can then divide the raw data by a master flat frame to produce an evenly lit final output. As with any other signal, flat frames bring their own noise, so it is important to average many flats to achieve a high signal-to-noise ratio.

Sensor non-uniformity

Each photosite is not uniformly sensitive to light. There is some variation in the charge created by a given number photons. The worst offenders are the hot (or cold) pixels discussed earlier, but there is a small amount of variation even between well-behaved photosites.

Further, there can be subtle vertical or horizontal patterns in the final image that are the result of the electronics used to read the sensor. While this pattern noise is typically a small fraction of the total noise, it is particularly bothersome because the human brain is designed to notice patterns. Most pattern noise can be corrected by flat frames, though patterns that shift between exposures can be particularly difficult to correct.

Uneven illumination

Illumination falloff around the outer edges of the field of view is an inherent property of all optical systems. Light coming through the system on the optical axis (through the center) will strike the sensor with greater density than light at the edges of a scene that enters at an angle. Optical systems can be corrected to varying degrees for optical vignetting, but some level of natural vignetting is unavoidable.

Any kind of physical obstruction in the light path also leads to uneven illumination. Dust motes close to the sensor are a common culprit, leaving diffraction shadows on the image. More distant mechanical obstructions like poorly designed internal baffles, lens spacers, or even over-sized dew shields can prevent even illumination across the field of view. This is why it is crucial that flat frames are taken with the exact same optical configuration as the light frames. When taking flats, make sure the camera is locked in the same orientation and the focus is the same. Some DSLRs have sensor cleaning functions that activate automatically on powering down, so make sure those are disabled, otherwise dust motes in the light frames will be moved or entirely gone in the flat frames.

Figure 16 is an example of a flat frame that has been enhanced to reveal the normally subtle differences in illumination. (The actual difference in brightness from the center to the edges was only about 15%, but the contrast was increased for illustrative purposes.) Both dust motes and vignetting are clearly apparent. We'll cover the details of proper flat fielding in the next part of the book.

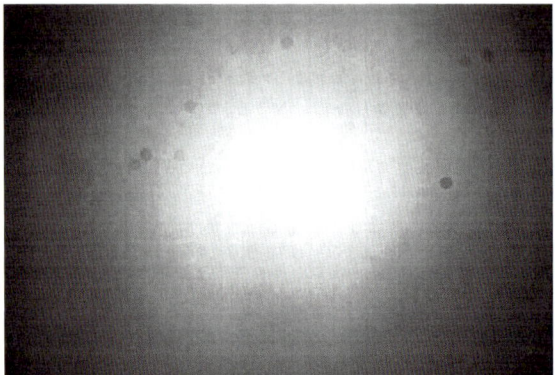

Figure 16. A flat frame (stretched to emphasize vignetting)

Adding signals and noise

All of the sources of noise in our images do not add together linearly. This section aims to visually illustrate how the various signals and their respective noise levels combine. Let's take a moment to quantify how signals and noise are quantified and see how calibration allows us to isolate the signal we want and reduce noise. The calibration process is covered fully in part three.

There are four sources contributing to the total number of electrons in each photosite:

1. The bias or offset current applied by the camera to every exposure during read out. This is a fixed current that is not time-dependent.
2. The dark current generated by thermal agitation of the silicon during the exposure.
3. The current generated by skyglow photons striking the sensor.
4. The current generated by target photons striking the sensor.

The challenge is that we are only interested in the fourth signal for our final image!

There are five potential sources of noise:

1. Read noise, which is inherent to the electronics in the read out process, and is independent of signal.
2. Shot noise associated with the thermal signal.
3. Shot noise associated with the skyglow signal.
4. Shot noise associated with the target signal.
5. Quantization noise, which reflects our inability to map one electron of charge to one ADU. It is usually insignificant, so we'll ignore it here.

In the subsequent steps of calibration and post-processing, we'll strive to improve the signal-to-noise ratio and subtract out the non-target signals, but for now we'll consider the example of a single exposure. In this case we have an exposure of a dim target taken under bright skyglow. At the end of an exposure, there are 586 electrons in the well: the bias level was 64 electrons, 300 electrons were generated by skyglow photons, 200 by target object photons, and 22 from dark current. The sensor's read noise is 15 electrons. These values are plotted in Figure 17. The error bars denote the noise levels.

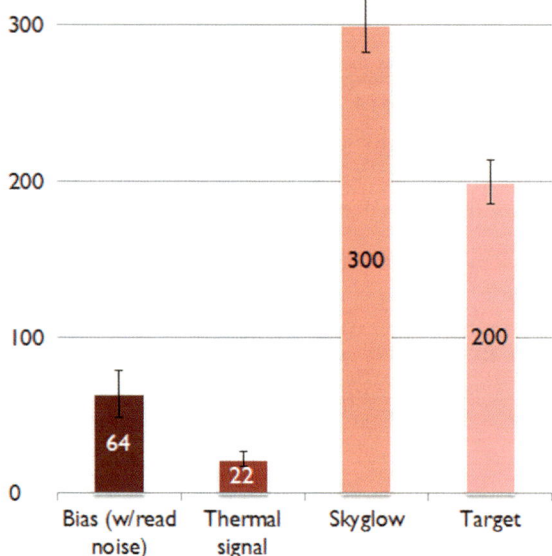

Figure 17. Four sources of electrons compose this exposure

Mathematically, we can add and subtract signals normally. Noise, however, is a little more complex. Since the measure we use for noise, the standard deviation, describes the variation around a mean, we can't simply add it like any other number. These variations can be positive or negative, so they can offset one another, and we have to account for that. To determine the total noise, we have to use a mathematical method called "summing in quadrature." This means that we square all of the noise values, sum those squares, then take the square root of the total.

$$Noise_{total} = \sqrt{Noise_a{}^2 + Noise_b{}^2 + Noise_b{}^2 + \cdots}$$

In this example, the thermal, skyglow, and target signals are all normal signals with their shot noise defined as the square root of each (4.7, 17.3, and 14.1 respectively). They sum to 522. As we've seen, there's no need for sophisticated math to determine the total shot noise; we can just take the square root of the total value: $\sqrt{522}$ = 22.8 electrons. The formula above will produce the same value:

$$\sqrt{4.7^2 + 17.3^2 + 14.1^2} = \sqrt{522} = 22.8$$

If we only had to deal with shot noise, we'd never need to think about summing in quadrature. We'd simply sum our signals and take the square root. But read noise is a different kind of noise that's not dependent on the total; it's a fixed value for every exposure. It is associated with the pseudo-signal of the bias current. While this is a fixed level of current applied to the photosites during readout, there is some variation in the precise number of electrons added each time. Combined with other electronic noise from the readout process, this amounts to the read noise.

In order to account for it in our noise total, we have to sum in quadrature:

$$\sqrt{15 + 4.7^2 + 17.3^2 + 14.1^2} = \sqrt{15 + 522} = 23.2$$

We'll later see how we can use calibration exposures to correct for each of these sources of noise. In addition to the flat frame that we've already seen, we'll use bias frames to quantify the offset current and read noise, and dark frames to quantify the thermal signal. Figure 18 shows how the values shown in Figure 17 compose a light frame, a dark frame, and a bias frame. The total noise for the light frame (the error bar) is +/- 23.2. For the dark frame, we have to add the shot noise from the thermal signal to the read noise:

$$\sqrt{15 + 22} = 6.1$$

And for the bias frame, there is no time-dependent signal, thus no shot noise to account for. The total noise is simply the read noise.

These analog charges are then quantified into digital values that can be stored in a file by the analog-to-digital converter. The third part of this book covers calibration and post-processing of these images. In calibration many light frames are averaged together, as are many dark frames, bias frames, and flat frames. These create a set of master frames, each with a far better SNR than a single exposure The master dark is subtracted from the lights to remove the thermal signal and offset value. The lights are also divided by the master flat to correct for light falloff.

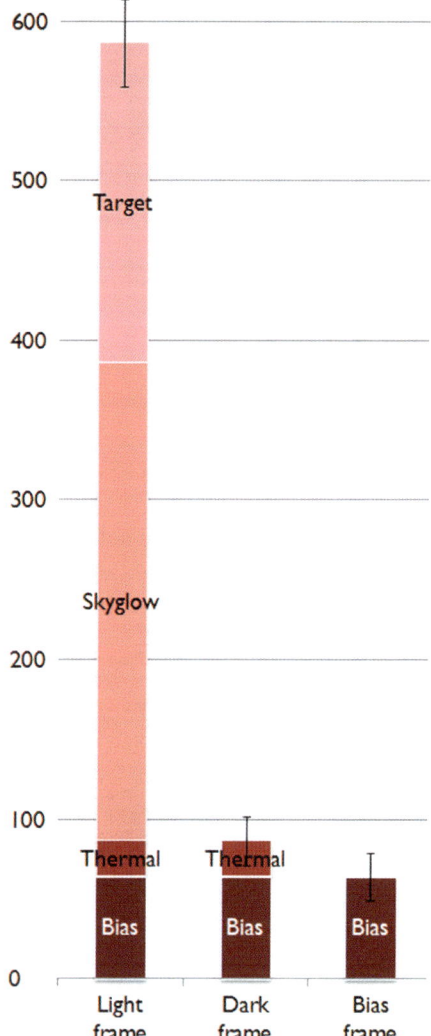

Figure 18. How these sources combine for each type of exposure

The calibrated image can then be aesthetically adjusted in post-processing. Part of that series of adjustments will include resetting the black point so that everything dimmer than the skyglow is set to black. This will leave only the signal from the target visible in the image, though all of the noise will remain. This again underscores the need for many exposures and careful calibration to control every source of noise.

Great images start with great data, so before we can start calibrating or processing any images, we need to capture sharp, clear, low-noise images for our raw material, which is the focus of part two.

Exercises

1.4 For the example light frame shown in Figure 18, what is the overall SNR?

1.5 Considering only the *target* signal, what is the SNR for this exposure?

1.6 What would the SNR be after averaging 10 exposures? How does that compare to the SNR of a single exposure that is 10 times as long (assume the three signals scale linearly)?

2 Aquiring Images

This section covers the details of image acquisition. You need the right equipment and techniques to collect quality image data, so we'll start with the mount, which is the crucial foundation of your imaging setup. Then we'll cover sensors, optics, and how to match the two properly.

To ensure your exposures are sharp, we'll review options for achieving optimal focus, as well as how to autoguide your mount with a computer. Then we cover the many accessories that are needed to bring your setup together. Narrowband imaging is introduced, as is the impact of the atmosphere on images. Finally, we establish a basic workflow for image acquisition.

IC 1396 through narrowband filters
(20-minute exposures through H-alpha, SII, and OIII filters - 13, 22, and 12 exposures respectively)

3   Mounts and alignment

Your mount is the stable foundation for your imaging system. Imaging demands far more from a mount than visual observing. Not only does it require greater stability of tracking from second to second, usually aided by autoguiding, but also a consistent (not rotated) field of view over many hours. Since an alt-azimuth mount's axes are not aligned with the sky's movement, it will suffer from what is called field rotation, where the field of view rotates over time, even though it remains centered on the same spot.

This is why the vast majority of imagers use a German equatorial mount (colloquially, GEM). Unlike a fork mount, a GEM's axes are aligned with the apparent rotation in the sky. To illustrate this, consider Figure 21 and Figure 22, which show the Big Dipper (or "Plough") asterism at three different times on one spring night as it rises in the northeast and rotates around the north celestial pole. (For bonus points, calculate the approximate latitude of this hypothetical observing site from the illustration.)

An alt-azimuth mount would have no trouble tracking a wide-field image centered on Megrez (the star where the handle meets the cup), but the telescope and sensor would not rotate as it is tracked. The result is that over time, the field of view will appear to turn around the point on which the mount is guided. Three images taken over the six-hour span in Figure 21 show this dramatically, but field rotation can be apparent even within a single exposure of a minute or less. There are field de-rotators available to correct for this by slowly turning the camera, but without them, your exposure times are very limited with this kind of mount.

The bottom axis of a German equatorial mount rotates with the sky, turning in right ascension (RA) to maintain the same field of view. Properly aligned, it tracks the rotation of the earth by turning on this axis once every 23h 56m. As you can see in Figure 22, the camera and telescope remain aligned with the target the whole night. The top axis of a GEM determines the declination at which the telescope is pointed. Turning the telescope on this axis always sweeps out a great circle that passes through both celestial poles.

In order to track accurately, a mount's right ascension axis (sometimes called the polar axis) must be pointed exactly at the celestial pole. Accomplishing this is called polar alignment. For visual observation or wide-angle imaging, a rough alignment within a degree or two will suffice. A polar alignment scope or other quick method will suffice in this case. For long exposure imaging, especially at longer

Figure 19. The Meade LX200 fork-style alt-az mount

Figure 20. The Celestron CGEM German equatorial mount

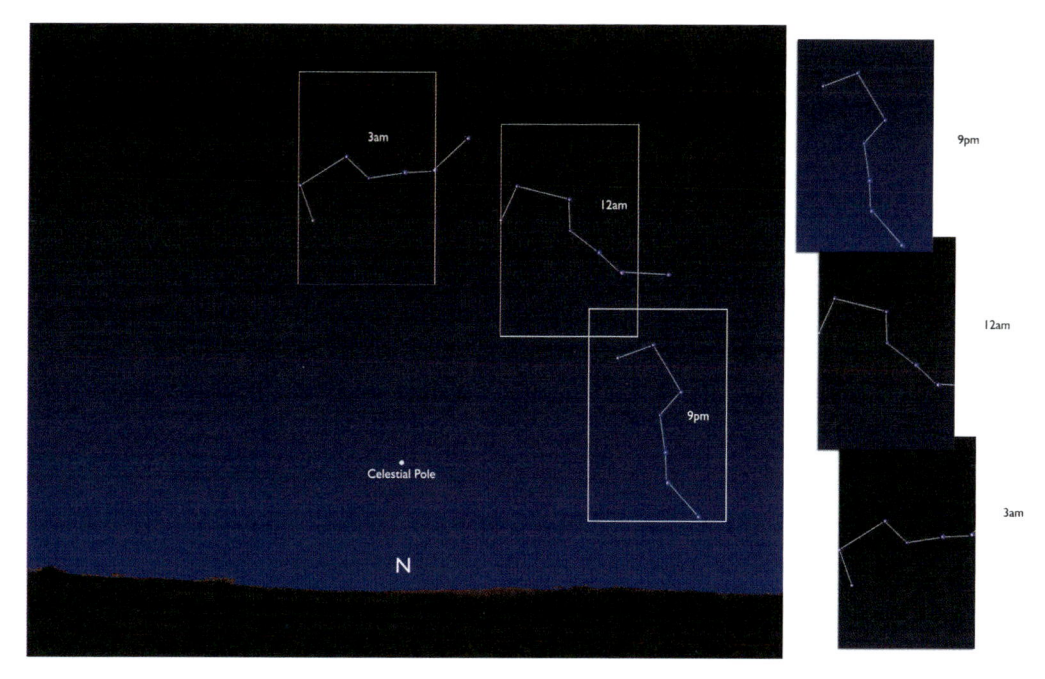

Figure 21. Field rotation between exposures taken on a fork mount

focal lengths, perfect alignment is critical, and that typically requires drift alignment or computer-assisted alignment, both covered later.

> Because the Earth travels 1/365.24th of its orbit around the sun every day, celestial objects rise about 4 minutes (1/365th of a day) earlier each evening.

Choosing and using a GEM

While it is tempting to first spend money on optics, a quality mount will do more to save time, effort, and frustration than any other component. Better-quality mounts typically have tighter mechanical tolerances that will produce tighter stars and sharper images. Their weight stands up to breezes better (even if it can lead to back pain). Some mounts provide automated polar alignment routines and more accurate go-to's. Buying a quality mount is like buying time—less wasted setup time and fewer wasted subexposures. If you only have limited opportunities to image, they are worth the additional cost.

The first time you use a GEM, it might seem difficult to point it where you want. Instead of simply moving up and down, left and right, the mount moves on the two rotational axes of right ascension and declination. It is not readily apparent which arrows on the hand controller will move the scope a particular direction in the sky, and the amount of motion isn't consistent between locations in the sky. Close to the pole, the RA axis motion translates to very little movement across the sky, while it only takes a little nudge to go a long way near the celestial equator. After a few nights out, however, the motions will become clearer, and the alignment process will get faster.

> In addition to creating some of the finest optics of his day, German optician Joseph von Fraunhofer (1787–1826) also invented the spectroscope, discovered the absorption lines in the solar spectrum that bear his name, and is credited with the invention of what we now call the German equatorial mount. Fraunhofer's insight in aligning an alt-azimuth's axes with the celestial pole allowed it to track the stars automatically by slowly driving one axis.

Meridian flip

One downside of equatorial mounts is that you must manage what is known as a meridian flip. With the telescope on the west side of the mount, it can follow an object from the eastern horizon up to the zenith. Any further motion to track the object westward will cause the scope to hit the mount, so the scope must be swung around to the east side of the mount, where it can follow the object from the zenith to the western horizon. This is known as a meridian flip.

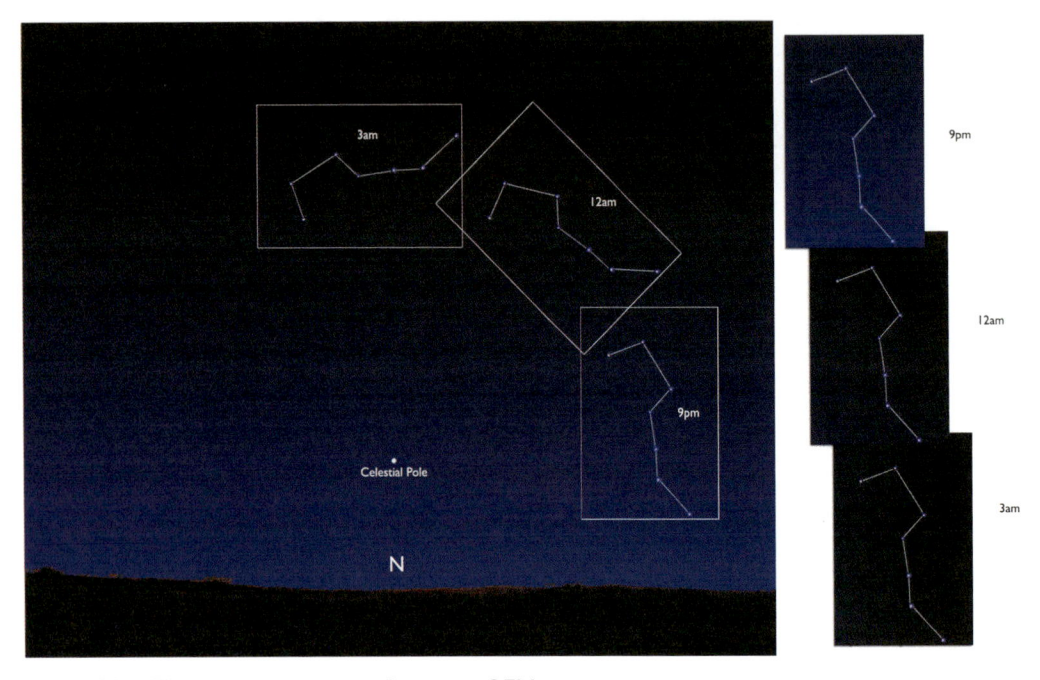

Figure 22. The same exposures taken on a GEM

The flip can be automated by some mounts, but it is typically done manually. There are two major consequences of a meridian flip for imagers, aside from the need to stay up late sometimes to manage it. The first is that the view is flipped 180° after the meridian flip. The second is that autoguiding needs to either be re-calibrated or notified that a flip has occurred, since the directions must now be reversed to achieve the same compensatory motion as before.

Polar alignment

It is important that an equatorial mount be polar aligned as accurately as possible. This means that the right ascension axis of the mount (the lower axis of the mount) is pointed exactly at the celestial pole. The sort of rough polar alignment that usually suffices for visual observation will lead to noticeable field rotation over the course of an imaging session. Even with short exposures, field rotation can show up if you are stacking many exposures of them, since there will be rotation between the first subexposure and the last. Most stacking software provides limited correction for this, but it's better to avoid it entirely. Unless you are shooting for only a short period of time or at very short focal lengths, good polar alignment is a necessity to keep stars sharp and round.

You can use a polar alignment scope to get the mount close to alignment, but it will take either drift alignment or a computer-assisted alignment procedure to achieve the level of accuracy needed for imaging. This can be time consuming, but as you get accustomed to the process and your equipment, it gets faster. One of the biggest advantages of having a home observatory or permanent pier is not having to polar align every time you set up.

Most recent mounts have a computer-assisted polar alignment routine that can be performed in only a few minutes. Under most circumstances, the alignment provided by this process is excellent, and long exposures can safely be taken on a mount aligned this way. The gold standard for attaining perfect alignment, though, remains drift alignment. While time-consuming even for the experienced, it will guarantee the accuracy of your alignment. Both of these methods require a way to line up a star in exactly the same place in the field of view. This was traditionally done with an illuminated cross-hair eyepiece, but the computer monitor or a DSLR's "live view" screen has obviated the need for that piece of equipment.

While autoguiding (covered in detail later) can compensate for most tracking errors, it does not eliminate the need for good polar alignment. Autoguiding compensates for small errors in tracking, but if the mount is not accurately aligned with the celestial pole, there will be field rotation in the image. Figure 23 is an example of poor polar alignment in an autoguided exposure of 30 minutes.

Acquiring Images

must have been just outside the upper left edge of the frame.

Tracking error

Even the best mounts with the best alignment have some imprecision in their ability to track the rotation of the sky. The primary source of this is periodic error, which is the error that comes from irregularities in the right ascension gears. This is seen as a periodic, sinusoidal error in tracking that repeats with every rotations of the mount's worm gear (usually every four to eight minutes). Computer controlled mounts can counteract this through Periodic Error Correction (PEC). To enable PEC, you typically enable an autoguiding setup to track a star, which allows the mount to see where it is running fast or slow. Once the mount's internal computer has "memorized" the periodic error, it can compensate for it from then on by speeding up or slowing down a little at the right times to compensate for imperfections in the right ascension gears.

If you plan to take unguided images, enabling PEC is a must, especially if using long focal length optics. If you

Figure 23. Field rotation from poor polar alignment

Because of the misalignment with celestial pole, the telescope is turning on an axis that doesn't quite match that of the sky. The autoguiding software is closely tracking the guide star, so stars near it remain round, but the field rotation becomes progressively more apparent further from it. You can see from the arcs of rotation that the guide star

Equipment Check — German Equatorial Mounts

Commercially-available equatorial mounts can be loosely categorized into three classes: budget, intermediate, and premium/heavy duty. For astrophotography, the rule of thumb is that your equipment should be no more than half the stated weight limit of the mount to be stably guided. This means that surprisingly large mounts are required to image with even modest telescopes.

The budget class mounts generally have stated visual weight limits of around 30 lbs (13.6 kg), which translates to a usable imaging limit of about 15 lbs (6.8 kg), depending on the length of the optics. Internally, most of them have machining tolerances that are fine for visual use and accurate go-tos, but are not optimal for long exposure imaging. Aftermarket modifications can smooth or replace some of the more problematic internal parts to improve performance, but they are still inherently less stable that larger mounts. Models will change from year to year, but in general, this class includes mounts like the Meade LXD-75 and LXD-55, the Celestron CG-5, Vixen Super Polaris, and Orion Sirius EQ-G. These are all clustered around the $600 (US) price point. A specialty GEM designed for travel, the Astrotrac, also has a weight limit in this class, but without computerized functions like go-to object seeking. While not the best choice for a permanent location, it is one of the few options specifically designed to be light and small, and it accomodates most standard photographic tripods.

The intermediate class of mounts can be a substantial step up in price (2–5 times as much as the budget class mounts), but they represent a substantial step up in quality as well. Most have visual weight limits around 40 lbs (18.1 kg) and feature much tighter internal machining tolerances that provide single-digit arcsecond peak-to-peak periodic errors. Mounts in the intermediate class currently include the Celestron CGEM and CGEM-DX, Meade LX80, Orion Atlas EQ-G, Astro-Physics Mach1GTO, iOptron iEQ45, and Losmandy's GM-8 and GM-11 (with the Losmandy and Astro-Physics mounts being a little harder to easily categorize than the other brands). For most people, intermediate mounts represent the limit of reasonable transportability and set up. Including counterweights, they can weigh up to 100 lbs (45 kg), which can make for a time-consuming set up in the backyard. This drives some people to establish semi-permanent piers, or use wheeled systems to position the mount.

The largest, sturdiest, and heaviest consumer mounts cover a huge range of sizes, but are generally big enough that you'll want to pier mount them or keep them in an observatory. The sky is the limit for the scopes they can swing around, and they carry price tags to match. Celestron's CGE and CGE Pro are probably at the smaller end of this class, with visual weight limits of 65 and 90 lbs (29.4 and 40.8 kg) respectively. Meade's LX800 also has a stated limit of 90 lbs, and it has additional features like a built-in autoguiding system. Other options include Software Bisque's Paramount MX and ME (with 90 and 150 lbs weight limits) and Astro-Physics' 900GTO, 1200GTO, and 3600GTO (70, 140, and 300 lbs). If you have a 20" telescope permanently mounted at a dark sky site, these are the mounts for you.

plan to autoguide your mount, it is better to leave PEC off. Autoguiding will correct for nearly all mount imprecision as it occurs, including periodic error, so there is no need for both systems to compete to adjust the mount at the same time.

Mounts also have random error that cannot be corrected by PEC. This usually comes from imprecision of the motors and control circuit. While autoguiding can correct low-frequency random errors, any quick or jittery movements should be corrected by fixing the inherent mechanical issue.

Drift alignment

The drift alignment process requires you to monitor the movement of a star in the field of view and iteratively make adjustments to the mount until it is polar aligned. It can be difficult to determine directions through the viewfinder or eyepiece, but it becomes quickly apparent once you start making adjustments. The real challenge of drift alignment is the time required and the tediousness of waiting several minutes for tiny amounts of drift. This can eat into precious imaging time, which makes the invention of computer-assisted alignment routines that take 5–10 minutes all the more important.

To drift align a mount, first monitor the magnified view of a bright star near the intersection of the meridian and the celestial equator. The meridian is the imaginary line that runs from north to south between the horizons. The celestial equator is the imaginary line halfway between the two celestial poles, running perpendicular to the meridian. If you are in the northern hemisphere, the region you are looking for can be found by tracing a line from Polaris, up through the zenith, to a point 90° away. For example if you are at 30°N latitude, the north celestial pole is 30° above the northern horizon, which means the highest point on the celestial equator (what you are looking for) is located 60° above the southern horizon. Once you have this southern star in your field of view, watch its motion on the north-south axis (ignore east-west drift for now). If it drifts south, the mount needs to be adjusted so that it points more to the west, and northward drift calls for the mount to be turned more to the east. This is usually accomplished by turning opposing azimuth knobs on the base of the mount. A magnified view of your camera's output (the back screen on a DSLR works nicely) can be used to establish a line on which the star can be monitored for drift. Once no drift can be seen even over the course of several minutes, the mount's azimuth is well-aligned.

To align the mount's altitude, swing the scope down toward either horizon along the celestial equator. Find a bright star around 20° above the horizon (not so close that atmospheric distortion will be an issue). As before, monitor the star's movement north or south from a line in the field of view. If you are on the eastern horizon, a southward drift means the mount needs to be adjusted upward and northward drift indicates a downward adjustment is needed. This is reversed if you are on the western horizon.

> Most mounts allow the scope to travel a few degrees past the meridian in either direction, so objects near there can be approached from either side. Set your mount to prefer approaching a target with the scope on the east side of the mount. This allows you to start following an object near the meridian before it transits, avoiding an unneccesary meridian flip.

4 CAMERAS

The camera is the next key piece of equipment to consider. There are many factors and trade-offs involved in choosing a camera, but as we'll see in the next section, the sensor must be considered in context with the optics; they should be chosen to function as a well-matched pair.

The first decision to make is whether to get a dedicated astronomical camera or a consumer DSLR. DSLRs (Digital Single-Lens Reflex) are mass-produced products, so they tend to offer the best value in terms of sensor size, and they can produce excellent results, but they were ultimately designed for daylight use. The scientific camera market is much smaller, so prices are higher, but these cameras are specifically designed with astronomical imaging in mind.

DSLRs

DSLRs offer large chip sizes, typically in the size range of APS film (about 16 × 24 mm), that would cost many times more in an astronomical CCD. A DSLR can also do double duty as an astronomical camera at night and as the family camera during the day. But aside from cost, DSLRs have other benefits as well. One of the most useful features of recent DSLRs is the ability to see the current view on the camera's rear screen. This allows for quick alignment and focusing right at the telescope without a laptop. Their low energy demand means that they can run on a small battery when necessary (though this is not recommended, especially in cold weather). While older DSLRs had some issues with read noise, most current DSLRs have levels rivaling those of dedicated astronomical cameras.

There are some limitations, of course. Since they have color filter arrays on the sensor, the potential for narrow band imaging is limited. With the exception of Canon's 20Da and 60Da models, they also have IR filters that block most the infrared and deep red portions of the spectrum. Aftermarket modifications can fix this, which dramatically improves the sensitivity to H-alpha objects, but at additional cost, voiding of the warranty, and difficulty using the camera for daylight photography. DSLRs also tend to have relatively small photosites, which makes them excellent matches for shorter focal lengths and wide field imaging, but not well-matched for long focal lengths. Unlike dedicated astronomical cameras, they do not generally offer photosite binning (where the output from multiple wells are read out as one to improve SNR but reduce resolution).

Canon currently seems to hold the hearts of most imagers using DSLRs. While they produced the only mass-market cameras designed for astronomy, the EOS 20Da and 60Da, their other cameras have become the workhorse cameras for many. Their main competitor, Nikon, offers a terrific range of lenses and accessories, and their cameras have very low noise levels, but there is significant in-camera processing applied to their "raw" image files, which makes the output far less useful for the calibration process necessary in astronomical imaging. Since calibration and stacking steps average many frames to reveal fine differences within the darkest fraction of the sensor's dynamic range, any such manipulation gets magnified. While there is a workaround for this problem, it is somewhat inconvenient. There may be some minimal processing applied to Canon's raw output, but the resulting files work very well for calibration.

Pentax has recently developed an astronomy-specific feature called Astrotracer that accomplishes limited sky tracking for wide field shots via the sensor electronics. That said, Pentax, Sony, and others have not reached critical mass in astronomical imaging, so most of the software and equipment has developed around Canon's product line.

Dedicated astronomical cameras

The range of available astronomical cameras is huge. Sizes range from tiny sensors of only a couple of millimeters square to huge sensors the size of medium format film. Many are monochrome, but some have color filter matrices for one-shot color imaging. Most have active cooling systems that hold dark noise to minimal levels by cooling the sensor well below the ambient temperature. There are clearly benefits to using products that were designed specifically for astronomical imaging.

Choosing a camera with a monochrome sensor gives you greater flexibility by allowing you to use any combination of filters, including narrowband filters, standard color filters, or clear "luminosity" filters. This flexibility does not come without burden, though. Filters and filter wheels can be very expensive, add weight on the focused, and there are additional learning curves involved in acquiring and processing the images.

The table summarizes the major trade-offs between DSLRs and dedicated cameras.

	DSLRs	Dedicated Astronomy Cameras
Active cooling	No	Usually
Near-IR sensitivity	Only with modification	Excellent
Cost per sensor area	Excellent	Generally high
Focus and align at the scope	Yes	Yes if laptop nearby
Narrowband imaging	Not well-suited	Yes (if monochrome)
Read noise	Low	Low
Thermal noise	High; requires dark temp. matching	Low with regulated cooling
Pixel size	Small (best for short f.l.)	A range available
Pixel binning	No	Yes
Can be used for other photography	Yes	No
Final image results	Great images can be produced by both	

Sensor size and pixel size

Figure 24 shows a comparison of the relative sizes of some common sensor sizes. "Full frame" is equivalent to 35 mm film, and a few cameras offer chips this size or even larger. The most common DSLR sensor size is the APS-C (approximately equivalent in size to APS film dimensions of 25.1 × 16.7 mm), which varies slightly by model. The Kodak KAF-8300 and KAI-2020 sensors are two popular sensors found in dedicated astrophotography cameras. The majority of astronomical sensors are branded by Sony or Kodak. Cameras from companies like SBIG, Orion, QHY, and FLI use these sensors, however the associated electronics that control them are usually proprietary.

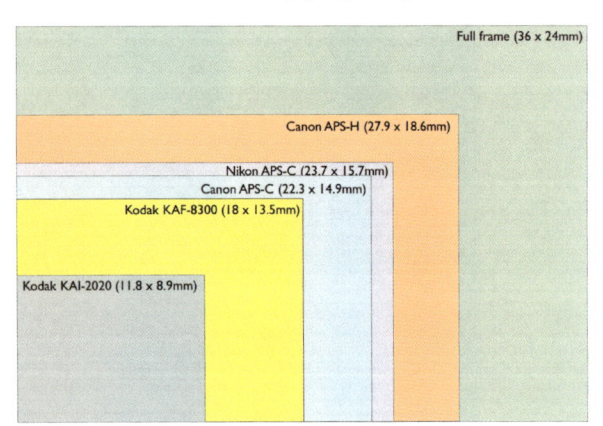

Figure 24. Common sensor sizes (to scale, but not actual size)

The race for more megapixels squeezed into the same sensor size in consumer cameras has led to progressively smaller pixels in DSLRs. While this means there are fewer photons striking each photosite, this has also been accompanied by corresponding improvements microlens technology to gather more light and lower read noise. More recent DSLRs exhibit excellent performance in most cases, but as we'll see in Chapter 6, their smaller pixels make them best suited for shorter focal lengths.

Connecting a camera to a telescope

Most telescopes have a standard 2-inch eyepiece opening, and this is where cameras typically connect for prime focus (not through an eyepiece) imaging. Unfortunately, cameras are fit for lenses, not eyepiece tubes, so various adapters are required.

DSLRs typically have a bayonet mount—familiar to anyone who has attached a lens where a quarter turn locks the lens with a click. To connect a DSLR to a telescope at prime focus, two adapters are typically needed. The first is a T-ring, which on one side has a bayonet mount designed to match a specific brand of camera, and on the other side has a standard "M42" threaded opening (42 mm with 0.75 mm thread pitch), sometimes called a "T-thread". The M42 side can then screw onto a generic T-adapter, which has male M42 threads on one side, and a metal tube that slides into the telescope like an eyepiece on the other. A T-ring can also screw onto a focal reducer or field flattened, many of which slide in like an eyepiece, but contain corrective lenses inside to shorten the effective focal length and/or reduce field curvature. There are also combined designs that are essentially T-adapters with a specific T-ring built as one solid piece (no M42 threads between them). Avoid any adapters smaller than two inches with DSLRs, since they will cause vignetting. Figure 25 and Figure 26 show one way a DSLR can connect to a telescope through a T-ring and field flattener. A T-adapter could be used in place of the field flattener.

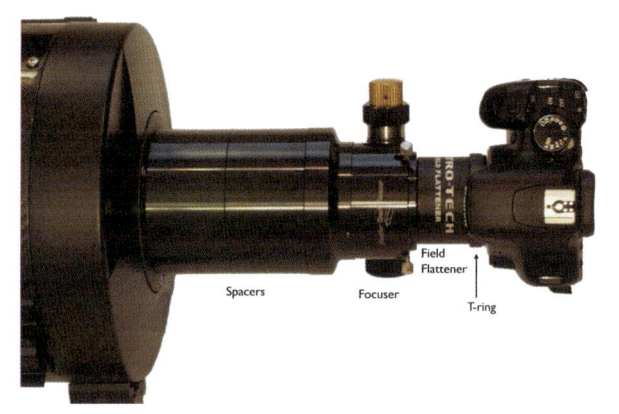

Figure 25. Typical connections between a DSLR and an 8" telescope

Figure 26. A view of the field flattener, T-ring, and DSLR separated

Most dedicated astronomical cameras connect via M42 or larger threads. Smaller CCDs may work fine with 1.25-inch adapters, but most sensors are now large enough to require 2-inch accessories. To make a solid connection that ensures the sensor is aligned with the light path, some focusers designed for imaging use only threaded connections. An example of such a rigid connection is shown in Figure 27.

Figure 27. Using only threaded adapters to attach a CCD camera

Bear in mind that the sensor must be a specified distance from any reducer or flattener, which may require precise spacers.

5   Optics

The telescope is probably the first thing most people consider in the purchase process, but for imaging it should always be thought of as part of an integrated package with the camera and mount. They should be considered together as a unit, taking into account what sort of objects will be imaged, the mount's capacity, and how it will match the camera's pixel size (covered in Chapter 6).

Any optical system is a compromise between many factors: aperture, focal length, focal ratio, flatness of field, optical aberrations, price, and ease of use, among other things. Finding the right one requires research and possibly some experimentation, but knowing the fundamental parameters that define every optical system will inform your decision.

Aperture is probably the most obvious parameter of a telescope. This determines the total amount of light gathered and maximum achievable resolution. In most reflecting optical systems, a portion of the total aperture is obstructed, which reduces the light gathered, but not the maximum resolution. Aperture is paramount for visual astronomy, but it is only part of the equation for imaging.

The other major parameter is focal length. This describes the distance from the last optical surface light passes through to the point of focus, but the practical implication for imaging is that focal length determines the field of view. Longer focal lengths capture narrower fields of view, and vice versa. You sensor sits at the point of focus capturing the light from this field of view. A short focal length instrument captures a wide field of view, thus each photosite on the sensor covers a larger area of sky than it would with a long focal length instrument attached. This description of how much area of the sky each pixel gathers light from is called the image scale. The focal length and the sensor's photosite size determine image scale.

Focal ratio is simply the ratio of focal length to aperture. Though it is derived from these two other values, it is the critical parameter for imaging because it uniquely describes the concentration of light per area on the sensor. (This excludes the effect of any central obstruction from a secondary mirror.)

Consider two instruments, both pointed at the constellation Orion: an f/8 telephoto lens with an 80 mm focal length (thus a 10 mm aperture), and an f/8 refractor with an 800 mm focal length and 100 mm aperture. Both are f/8, but the telescope has an aperture with ten-times the diameter (and thus on hundred-times the light gathering surface area) as the telephoto lens.

Using the same camera, we take a one-minute exposure through each optical system. The image taken through the telephoto lens will show a very wide field. Depending on the size of our sensor, it may even show the whole constellation. In such an image the Great Orion Nebula (M42) is a pink smudge in the hunter's belt, only a few dozen pixels across. The image taken through the telescope will show a much narrower field of view because the focal length is ten times longer. This time, the wispy nebulosity of M42 fills much of the frame, revealing fine details in its structure. But if you were to measure the brightness of the nebula's core in each exposure, they would be the same. Counter intuitively, the greater aperture of the 100 mm scope does not yield a brighter image because the focal ratio is the same.

Consider that the telephoto lens is gathering all of the light from a huge area of the sky, and then bringing it to focus on the sensor. The longer focal length of the telescope means that it gathers light from a much smaller area of the sky. This smaller amount of light has to cover the same sensor, so the only way to keep the light flux per pixel constant is to increase the aperture. Thus, to keep the same exposure time, the aperture has to grow as the focal length increases, and that means a constant focal ratio. Greater focal length defines a narrower field of view, and greater aperture provides better resolution, but for a given focal ratio, the brightness of an extended (non-point source) object will be constant.

For over a century, photographers have described their lenses in terms of focal length and focal ratio. The same rationale applies to imaging with telescopes. While the three parameters are intertwined, the focal length and focal ratio determine the key parameters for imaging—size of the field and brightness at the sensor.

In the same way photographers use lenses of different focal lengths for different subjects, different telescopes and camera lenses are required to capture the range of deep-sky targets, from widefield images of the Milky Way down to the smallest galaxies and planetary nebulae. (When you explain to your significant other why you "really need this new telescope," the science of optics is on your side!)

Telescopes for visual vs. imaging use

The qualities that make an excellent visual telescope are different from those required for imaging. The compromises inherent in any optical design make it very difficult to produce one telescope that excels at both. For instance, the focal ratio is of little importance for visual observers, where brightness is influenced by the exit pupil, and various eyepieces can be used to obtain the desired magnification and exit pupil. Imaging telescopes demand faster focal ratios that concentrate the light for shorter exposure times and better SNR. They also require flat fields that put the entire sensor in the critical focus plane. The entire sensor should be fully illuminated, which requires wide optical stops and large focusers. Finally, chromatic aberration must be controlled across a wider range of wavelengths, since electronic sensors are sensitive to 'colors' just outside of the range of human vision.

Optical aberrations

A theoretically perfect optical system would bring all light rays from across the entire field of view at all wavelengths to convergence across the full width of the sensor. The ways an optical system diverges from this perfection are known as aberrations. Optical designers must make trade-offs between which aberrations to control, the feasibility of manufacturing, and practical use issues like weight, length, focal ratio, etc. Before considering the major telescope designs, it's helpful to understand the types of aberrations.

Even with perfect optics, the wave nature of light introduces diffraction effects that cannot be overcome. We'll cover these in detail in Chapter 6, but it's important at this point to know that light from a point source of light like a star is spread by diffraction into a point known as the Airy disk. An optical system whose aberrations are controlled to the point where they are smaller than the Airy disk are said to be diffraction-limited. This is as good as it gets; you can't overcome the effects of diffraction. Note however, that an optical system may be diffraction-limited across only the center of its field of view, or only along a curved focal plane, so the term does not necessarily describe all possible aberrations. Systems whose aberrations are larger than the Airy disk are known as aberration-limited.

In general slower telescopes (higher focal ratios) control aberrations better for a given design. Creating a large, flat, diffraction-limited field with a fast focal ratio requires more optical elements, introducing manufacturing complexity and the potential for secondary aberrations. Such telescopes are highly prized by imagers, and they are accordingly expensive.

There are six primary types of optical aberrations imagers should be aware of when considering a telescope:

Chromatic aberration is where different colors of light come to focus at different distances. This is a result of the fact that lenses refract shorter wavelengths of light more than longer wavelengths. Figure 28 shows the simplest telescope design, a singlet refractor, which shows substantial chromatic aberration. Mirrors do not suffer from chromatic aberration, but any lens element in a compound system can introduce it. Longitudinal chromatic aberration causes a symmetrical colored halo around bright stars across the field of view, since part of the spectrum is out of focus. Lateral chromatic aberration is where stars are stretched into linear spectra that grow larger away from the center of the field.

Coma is an optical aberration where off-axis (away from the optical center) stars appear not as points, but as wedge- or comet-shaped flares that point toward the center. Refractors do not suffer from coma, but it is the limiting aberration of Newtonians and many Cassegrain designs. Coma the result of light from off-axis angles coming to focus at different distances depending on where it strikes the mirror, thus coma gets worse as you move away from the center of the field. It can be partially corrected with lens systems aptly known as coma correctors. Coma is dependent on focal ratio, with faster reflectors suffering more.

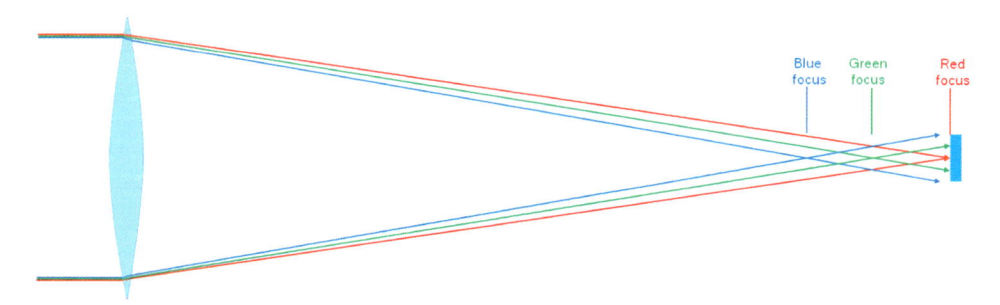

Figure 28. A singlet refractor exhibits substantial chromatic aberration

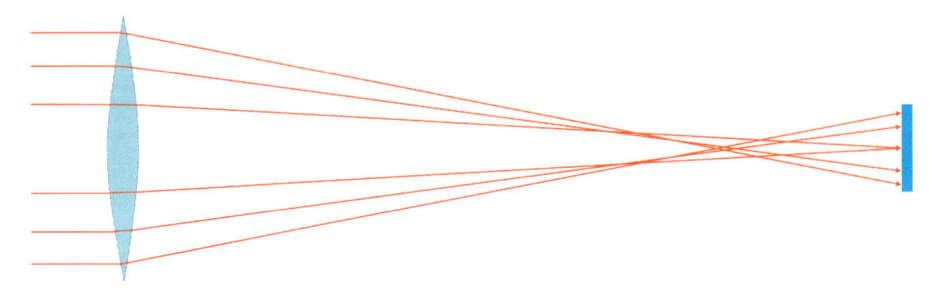

Figure 29. Spherical aberration

Field curvature is where the focal plane is not flat. The specific curve of a system is known as its Petzval curvature. In general, light arriving from off-axis comes to focus closer than on-axis light. This creates a focal plane that is curved like the inside surface of a sphere. The human eye corrects for this when observing visually, but sensors are flat. Thus if the center of the field is in focus for imaging, the edges are not, and vice versa. In some cases, there may be a compromise focus in the middle that is "good enough" across the whole field of view. This good-enough focus may actually be diffraction limited for some systems. Field flatteners are designed to correct field curvature, and since this aberration is usually determined by the focal ratio, flatteners are designed to work for specific ratios.

Astigmatism is generally only seen in Ritchey-Chrétien designs, and its appearance is similar to that of coma except stars are stretched into ovals rather than comets. Like coma, it is more pronounced toward the edges of an image. Astigmatism is related to field curvature, and field flatteners can provide some correction to astigmatism in R-Cs.

Spherical aberration is where light rays entering the system from the edge come to focus at point closer than those from the center (Figure 29). Both refracting and reflecting designs are susceptible to this aberration. This affects the entire field of view, and there is no compromise focus--no matter how you turn the knob, the image never comes into sharp focus. Because of this, it is perhaps the most important aberration that telescope designers control for, and it is almost never a limiting aberration for modern optics. In reflecting systems, parabolic mirrors are free of spherical aberration, but spherical or hyperboloid mirrors must correct for it. In refractors, multiple lens elements can minimize spherical aberration by using offsetting lens curvatures. A related secondary optical aberration, spherochromatism, is where spherical aberration varies with the wavelength of light. The result is similar to chromatic aberration.

Distortion is most noticeable in daytime photography, for instance when taking a picture of a brick wall and the straight lines of the grid looked curved. As long as it is not extreme, distortion is not usually much of a concern in astronomical imaging, with the exception of mosaics, where multiple images are stitched together to show a wide field of view. Even a small amount of distortion creates a challenge for the creation of a mosaic.

Refractors

Refractors were the first telescopes, and they remain some of the finest. The use of large refractors in scientific astronomy declined in the early twentieth century because of their weight, but small refractors remain a favorite with amateurs, especially in the imaging community. The most basic refractor is a single-lens system we saw in Figure 28. While this schematic illustration is exaggerated for effect,

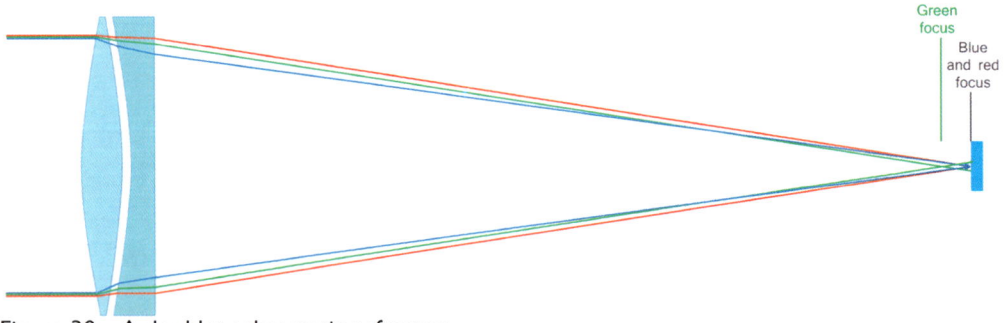

Figure 30. A doublet achromatic refractor

you can clearly see that there is a focus problem when you put a sensor behind it. Light of different wavelengths is refracted at different angles as it passes through a lens, which leads to convergence at different distances from the sensor. This chromatic aberration results in color fringes or halos around bright objects. Early refractors controlled this false color through extremely long focal ratios, sometimes greater than f/100.

The chromatic aberration problem was partially solved in the 18th century through the use of multiple lens elements with different diffraction indices. There are many arrangements of lenses used (and many associated patents), but the idea is to bring a wide range of wavelengths to focus in the same plane. While it is not feasible to bring all wavelengths of light to exactly the same focal point with lenses, these multiple element systems can bring them close enough together that the in-focus spot sizes are smaller than the Airy disk.

Achromats were first developed around 1760, using two lens elements with different refraction indices (Figure 30). A low dispersion "crown" glass lens is paired with a higher dispersion "flint" glass lens. This design brings the red and blue ends of the spectrum to simultaneous focus, while the middle (green) region of the spectrum focuses in front of the red-blue focus plane. Visually, the result is far less false color than a singlet, which allowed for shorter telescopes to be developed now that extremely high focal ratios weren't required to control color. The use of modern ED (Extra-low Dispersion) glass and computer-optimized designs allows for far better color correction than these original doublet achromats. A hypothetical focus-shift diagram for this achromatic telescope is shown in Figure 32.

It took advances in glass making at the end of the 19th century before the next step in refracting telescope technology could be achieved: apochromatic optics. By definition, apochromats bring three wavelengths to focus, typically using three lens elements, though doublet apochromats are possible. All modern apochromats use ED glass in at least one element. Figure 31 shows a schematic of a triplet lens system. True apochromats exhibit no false color around bright objects.

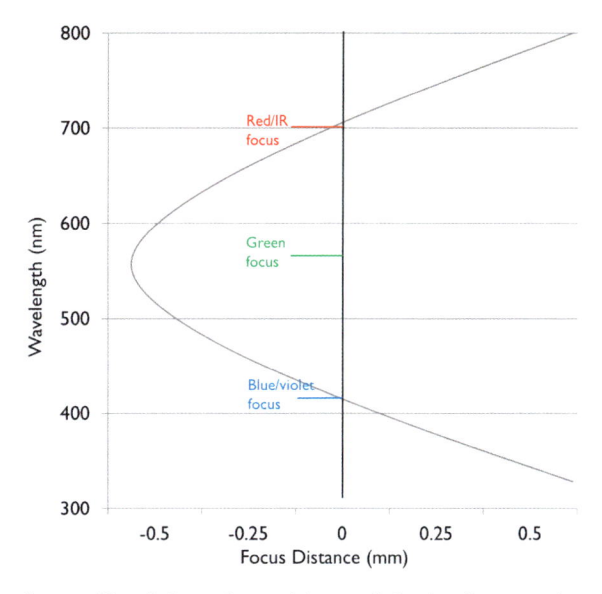

Figure 32. A hypothetical focus shift plot for an achromatic system

There are numerous ED glass types used by manufacturers. You cannot judge the performance of a refractor simply by the number of lens elements or the type of glass. The lens elements work together, so even if one is of the rarest glass, if the others are not designed to complement it or the elements are not accurately figured or aligned, image quality will suffer. Some long focal length achromatic doublets perform exceptionally well, especially for planetary viewing, and some of the best ED doublets can outperform lower quality triplets. In general though, a triplet of equal quality will substantially outperform a doublet.

A hypothetical focus-shift diagram for an apochromatic telescope is shown in Figure 33. In addition to bringing

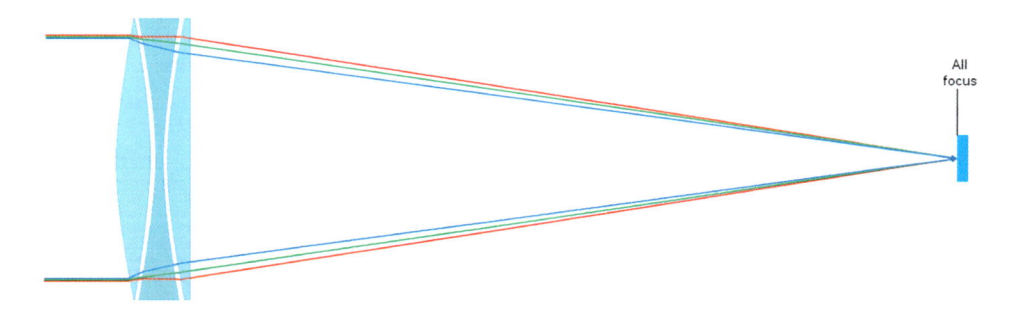

Figure 31. A triplet apochromatic refractor

three wavelengths into focus, the apochromat also has a narrower overall spread of focus within the visual spectrum than the achromat—the wavelengths that are most out of focus are not as far out of focus. They are in a close enough range to bring most of the visible spectrum into sharp focus in the same plane.

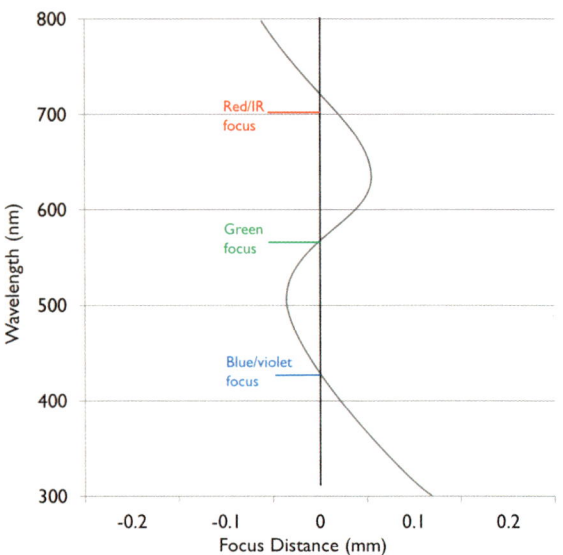

Figure 33. A hypothetical focus shift plot for an apochromatic system

Color correction is not the only attribute on which to judge a telescope for imaging, however. Even if color is perfectly controlled, other aberrations can affect imaging performance. For instance, all refractors with only objective lenses, no matter how many are used in combination, will have some curvature of field. Since electronic sensors are flat, a field flattener is needed to ensure that the edges of the field of view remain in focus. Even better than adding a field flattener, some refractor designs produce flat fields by design. Quadruplet or Petzval-type designs use an ED doublet objective paired with a two more lenses closer to the sensor that control for color and flatness of field.

Refractors offer some of the sharpest views of all telescope designs, and doublet or triplet refractors are an excellent choice for astroimaging. They are easy to use, and they almost never require collimation or maintenance. If you also plan to do visual observing, the high contrast, tack-sharp views through a good refractor are hard to beat. All of this comes at a price, however. Quality refractors are among the most expensive telescopes per inch of aperture. Because of their weight and length, apertures larger than about five inches require premium-class mounts. Refractors larger than six inches are getting into "yard cannon" territory.

> Two of the surviving telescopes used by Galileo have focal lengths of about 980 mm and 1330 mm. These telescopes had objective lenses of 37 mm and 51 mm respectively, but both were significantly stopped down, to 15 mm and 26 mm, yielding effective focal ratios of f/65 and f/51! With the eyepiece lenses Galileo used, they had fields of view only about a quarter of a degree across.

Telephoto lenses

A good telephoto lens can be used to image the largest deep-sky objects, and even a basic wide-angle lens can capture beautiful wide-field shots that can allow the incorporation of the landscape into the image. When choosing a lens though, quality is critical. Daylight photography is far more forgiving of optical defects, and most lenses are designed with this in mind. Chromatic aberrations, field distortion, and astigmatism that would never be apparent under normal photographic circumstances are all too easily revealed by the flat fields and pinpoint stars of the night sky.

There are a couple of things to keep in mind if you are using a camera lens. The first is that prime lenses (fixed focal length) nearly always perform better than a zoom lens of equivalent quality. Zoom lenses require more glass elements and design compromises to keep them compact. Prime lenses will generally provide a sharper image with less distortion, and some excellent manual focus primes are available on the used market. The unforgiving nature of astronomical imaging requires most lenses to stop down a little from their widest aperture. Normally we want to gather as much light as possible, but narrowing the aperture one or two stops will yield a sharper image, even if it requires more exposure time. Distorted stars will ruin the appearance of any night sky image, so it's important to stop down enough that their appearance is reduced.

Also be careful of older lenses designed for film photography. Most are not designed to control chromatic aberration across the wider range of spectral sensitivity that electronic sensors offer, so they can cause "blue bloat" around stars as a result of UV light that is not brought to sharp focus.

There are many advantages to using standard camera lenses. The weight of short focal length lenses is inconsequential to a standard telescope mount, so minimal balancing is required. The shorter focal length is more forgiving of tracking error, so lesser quality mounts work well, and autoguiding is rarely required. Many options are available to mount a camera to the common Losmandy or Vixen-style dovetail plates using the camera's tripod mount. For those

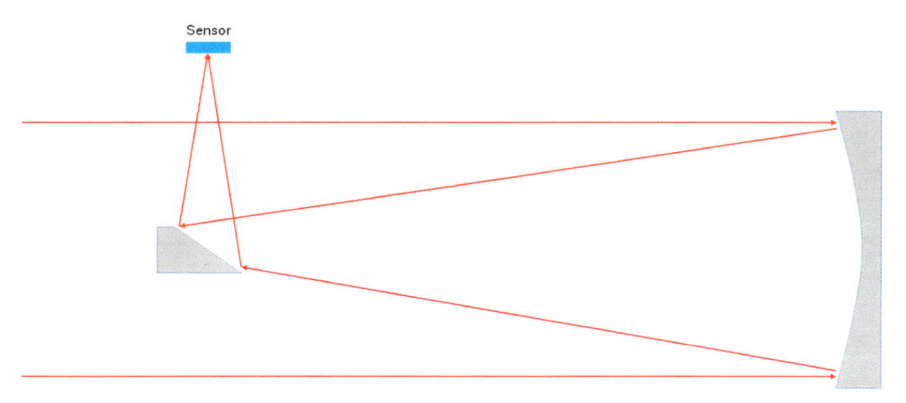

Figure 34. The Newtonian design was the first and most basic reflector

interested in a do-it-yourself project, a 1/4"-20 bolt and some materials from the hardware store can quickly be assembled into a stable adapter. Longer lenses can require support at the camera and also on the lens.

Astronomical telescopes are generally cheaper than premium camera lenses above 500 mm in focal length, though some excellent telephoto lenses rival the best apochromatic refractors. But for wide-field images that require less than 500 mm, camera lenses are by far the more plentiful option.

Reflectors and compound telescopes

Reflecting telescopes use mirrors to gather light instead of lenses. They can be manufactured at sizes that refractors can only dream of, and mirrors bring all wavelengths of light to the same focus, so they do not suffer from chromatic aberration. Compound telescopes combine mirrors and lenses; usually a large mirror to gather light with lenses used to correct the view. However, both have their own set of drawbacks and aberrations that can vary significantly between the designs.

Newtonians (Figure 34) are the original reflecting telescopes. As their name implies, they were invented by Isaac Newton in 1668, and their parabolic or spherical primary mirror surfaces are the simplest to manufacture and a favorite of amateur telescope makers. The secondary mirror is usually held in place by thin 'spider' vanes. Dobsonian telescopes are Newtonians on a simple alt-azimuth mount, making them ill-suited to imaging.

Newtonians offer the best value per amount of aperture, but they do have some drawbacks for imaging. Even perfectly crafted parabolic mirrors suffer from coma, which spreads the light from a point into a conic shape, especially toward the edges of the field. This can mostly be corrected by coma corrector lenses, which are necessary for imaging. Newtonians' size also makes them difficult to mount equatorially, which is a drawback for imaging. On some Newtonians designed for visual use, there may not be enough inward travel on the focuser to bring an image to focus on a camera. Finally, Newtonians require careful maintenance of collimation to perform well.

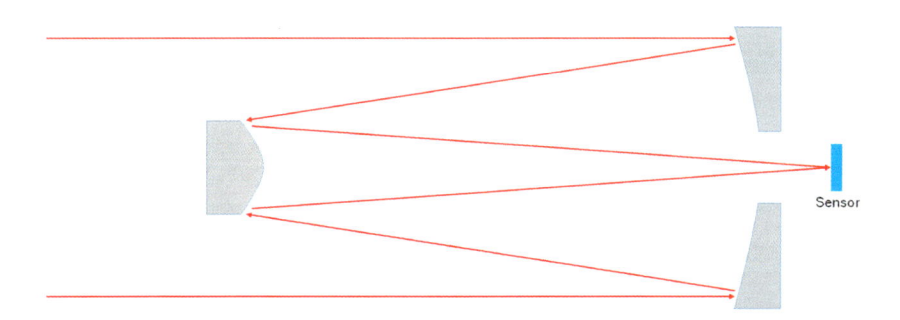

Figure 35. The Ritchey-Chrétien design

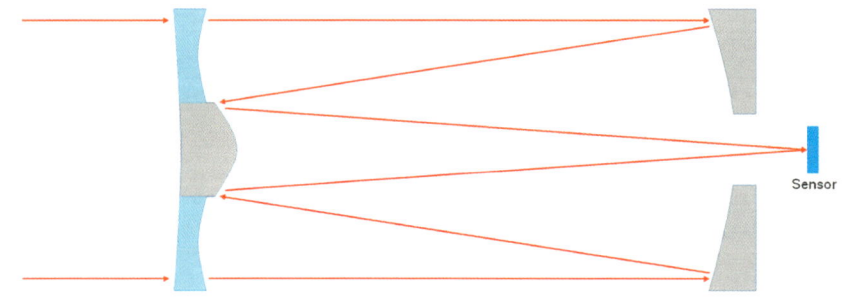

Figure 36. The Schmidt-Cassegrain design

All of these are bugaboos are greatly reduced if you purchase a Newtonian that is designed for imaging. There are 8- to 10-inch imaging Newtonians that offer an exceptional value as long as your mount can handle their weight. The best incorporate Wynne coma corrector optics that are specifically designed for that scope's primary mirror.

The Ritchey-Chrétien telescope design (Figure 35) used to be seen only in large professional telescopes, but recent improvements in manufacturing hyperbolic mirrors have made them available in amateur sizes and prices. Like Newtonians, R-C's have an open tube design with no corrector plate. Spider vanes hold a large secondary mirror, so brighter stars will produce diffraction spikes. Also like Newtonians, they are pure reflectors, with no lens elements. Unlike Newtonians though, R-C's are a Cassegrain design: the secondary mirror is curved, rather than flat. In particular, they use hyperbolic mirrors for both the primary and secondary mirrors, which eliminates coma, though it introduces some astigmatism. Most smaller R-C's are also symmetrical designs, where the image is brought to focus at the back of the telescope through a hole in the primary mirror.

Ritchey-Chrétiens are excellent medium or long focal length imaging telescopes, and their folded design means their tubes are shorter than Newtonians. They are not typically available in focal ratios as fast as Newtonians, which can be in the f/4 range, but their design and coma-free optics are a good fit for imaging. The primary mirror is typically fixed, which means that collimation is rarely required.

R-C's do suffer from field curvature and related astigmatism, both of which can be corrected with a field flattener. They are primarily designed for imaging—visual use of a Ritchey-Chrétien is usually disappointing due to the low contrast views that result from their large central obstruction, but this same central obstruction makes them ideal for imaging by minimizing vignetting.

For decades, since Celestron's introduction of the C8 in 1970, Schmidt-Cassegrain telescopes (SCTs) have been the workhorses of amateur astronomy. Their short-tube design and value have made them the most popular telescopes in the eight- to sixteen-inch range. SCTs are catadioptric telescopes, meaning that they use both a mirror and a lens in their design (Figure 36). The thin corrector plate that serves as the lens also seals the tube, keeping dust and dew off

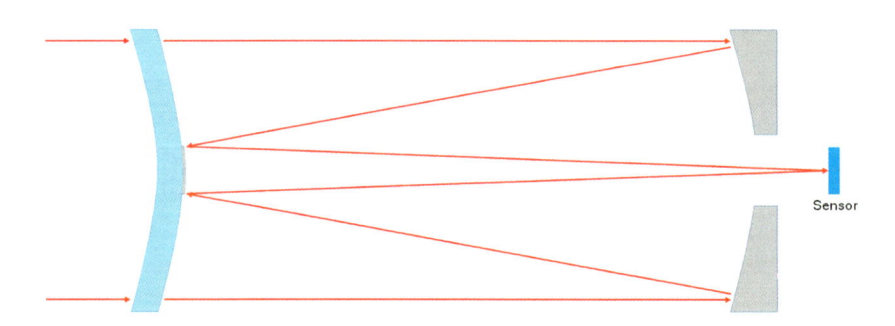

Figure 37. The Maksutov-Cassegrain design

of the primary mirror; however, this leads to longer cool-down times.

SCTs are focused by moving the primary mirror, but most models now include a lockable mirror for imaging. Fine focusing can be achieved by moving the camera instead, usually with a Crayford-style focuser. Be sure that your SCT has a lockable mirror if you plan to use it for imaging. Unsecured SCT mirrors can settle or shift a little as the mount tracks and gravity pulls the mirror at a different angle than when it was focused. This can lead to out-of-focus images when the mirror shifts (or 'flops') in the middle of an imaging run.

While the most common mount for SCTs are alt-azimuth fork-mounts, they can also be mounted on equatorial mounts. Imaging with a fork mount is possible with a wedge to align the altitude with the celestial north pole and a field de-rotator, but for many people, it's simpler to get an equatorially mounted SCT.

SCTs are typically around f/10, using primary mirrors of around f/2 with secondary convex mirrors of about f/5. The focal plane is curved, leading to out of focus stars on the periphery of the image, but this is correctable with a field flattener. Focal reducers can be used as well, most yielding an f/6.3 system. Schmidt-Cassegrains are also the only telescopes that can take advantage of the Hyperstar or Fastar imaging systems, from Starizona and Celestron, respectively. These products allow the user to mount a camera in place of the secondary mirror, yielding an incredibly fast f/2 system. They are not without drawbacks, however. Filters cannot typically be accommodated for narrowband imaging, and focusing is more tedious, since the sensor must be aligned parallel to the mirror.

Schmidt-Cassegrains are good all-around performers, both visually and for imaging. As long as a focal reducer is used, nearly all SCTs will perform reasonably well for most imaging purposes. Their focal lengths are not for wide-field shots, though with the Hyperstar or Fastar systems, one can accomplish this. Like Newtonians, SCTs typically suffer from coma, but newer "coma-free" designs rival the performance of Ritchey-Chrétiens. These recent improvements to the SCT design significantly improve performance for imaging. Meade's Advanced Coma-Free (ACF) series and Celestron's EdgeHD both correct for the coma inherent in the SCT design. The EdgeHD scopes also add a field flattener to also correct for field curvature.

Like SCTs, Maksutov-Cassegrains (Figure 37) are catadioptric telescopes, but they have a thicker meniscus lens in front, with a small secondary mirror on the back of the lens. The view through some MCTs is as sharp as an apochromatic refractor. They are nearly free of coma and chromatic aberration, and they have a small central obstruction. The primary issue with MCTs is their long focal length: they are typically f/15. Like refractors, it is also difficult to get large aperture Maksutovs. The largest commercially made MCTs have apertures of only 150 mm or 180 mm. The meniscus lens is large and heavy, which makes it difficult to manufacture, but also gives the telescope enormous thermal inertia. Cool-down can be very long, and most models over six inches in aperture have fans to assist in achieving thermal equilibrium. Despite these caveats, if you have a mount that can handle one, they can be an excellent choice for imaging small targets, like planets, planetary nebulae, and small galaxies.

Two other less common designs are the Maksutov-Newtonian and the Schmidt-Newtonian. Both are variations on the Newtonian, but with better correction for coma and other aberrations. Maksutov-Newtonians (Figure 39) have significantly reduced aberrations compared to a simple Newtonian, with focal ratios that are typically around f/6, making them much faster than their MCT cousins. The Intes Micro MN series of Mak-Newts are very well-regarded examples of this design. Schmidt-Newtonians combine

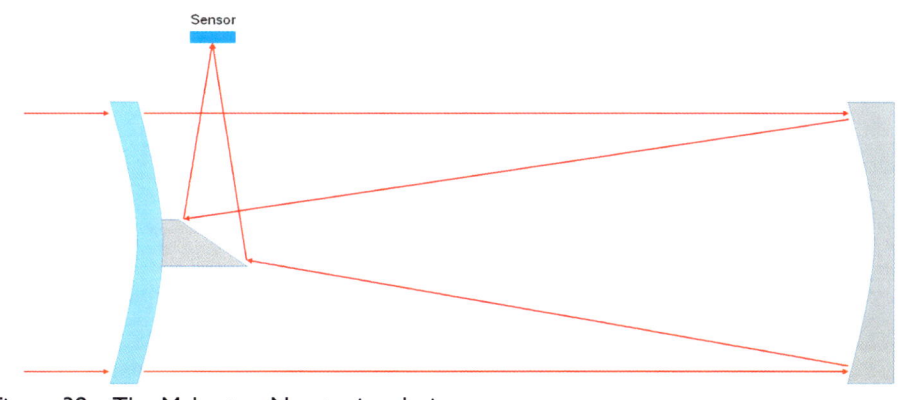

Figure 38. The Maksutov-Newtonian design

Equipment Check	Brand ecosystems

While some accessories fit universally, buying a particular brand of telescope can commit you to a particular "ecosystem" of adapters and accessories. At some point down the road, you may find yourself buying an adapter for an adapter or a spacer ring that costs as much as an eyepiece, and think, "How did I end up sinking so much money into this?" It's a place that many imagers find themselves, especially as they try to accomodate fitting multiple cameras with different telescopes.

The highest quality telescopes seem to have the most proprietary accessories. There are no doubt many benefits to this: machine tolerances are exceptional, you can be assured that the required optical distances are as specified, and threaded adapters hold a camera parallel to the image plane much more accurately and rigidly than a standard two-inch eyepiece holder. Takahashi, Televue, and Borg telescopes all have specific imaging parts, with the requisite charts, diagrams, and part numbers to guide you to imaging paradise. (With their limitless modularity, an argument could be made that Borgs are composed entirely of accessories.)

Aside from a self-deprecating sense of humor, the only "cure" is to know what you are getting into at the beginning. Acknowledge that the price of the OTA (optical tube assembly) is just the beginning. If it's going to need a better focuser, consider that part of the up-front cost. Research the price of the required adapters and spacers to fit your camera so there are no surprises. Even better, find a kind soul online to sell you a complete package with all the upgrades and accessories you need.

the corrector plate of a Schmidt-Cassegrain with the Newtonian design. This reduces coma and allows for faster focal ratios than most SCTs.

Telescope quality

Aside from the optical design of a telescope, always consider the construction. Quality telescopes are made with quality materials. The tubes are well-baffled inside to limit stray light. The rings and dovetails are sturdy. And for imaging, a good focuser is critical. Imaging requires a focuser that is strong enough to handle the weight of a camera without slipping and can be locked without altering the focus. Few standard focusers meet these criteria, and as a result, there is a thriving after-market focuser industry with imagers in mind.

> Any telescope that uses straight spider vanes to hold the secondary mirror will produce diffraction spikes around brighter stars. This can be aesthetically pleasing, but combining images from multiple nights where the camera is not in exactly the same position will result in misaligned spikes. Curved vanes do not produce spikes, and after-market replacements for straight vanes are available for many scopes.

6 IMAGE SCALE: MATCHING SENSOR, AND OPTICS

Resolution and seeing

Because of their incredible distances from us, all nighttime stars can be considered point sources of light. Even at perfect focus, the light from a star cannot be resolved to a point because of the limits imposed by diffraction. The disk you see as you increase the magnification is not a close-up view of the star, but an interference pattern called the Airy disk. For an ideal telescope with no central obstruction, the Airy disk is a bright disk containing about 84% of the total light, surrounded by several progressively dimmer diffraction rings (known together as the Airy pattern), shown in Figure 40. Central obstructions from the secondary mirror of reflecting and compound telescopes increase the proportion of energy in the outer rings. Optics where aberrations are controlled enough that the limiting resolution is due to the wave-like nature of light revealed by the Airy disk are described as diffraction-limited.

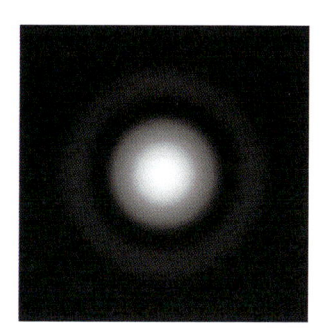

Figure 40. The Airy disk (simulated)

The diffraction pattern is not named for its diffuse appearance. It is named for Sir George Biddell Airy, the English astronomer and mathematician who first explained it in 1835. Airy also made significant contributions to planetary astronomy and engineering mathematics, but this disk of starlight immortalized him among astronomers.

The size of the Airy disk represents the resolving limit for an optical system, since any smaller detail is lost in this diffraction pattern. Since the size of the Airy disk determines resolution, it's useful to be able to calculate it. The formula for the spatial size of the central disk, between the first minimum on each side, at perfect focus is

$$Width \cong 2.440\lambda \frac{f}{a}$$

where λ is the wavelength of light, f is the focal length, and a is the aperture. You can see that f/a in this equation is the focal ratio, which means that the size of the central disk is dependent only on the wavelength of light and focal ratio of the optics. Thus all telescopes of the same f-ratio, regardless of the size of their aperture, project Airy disks of the same *physical* size at focus. For instance, all f/5 optics will show a point source of green (555 nm) light as an Airy disk with a radius of 3.39 μm; all f/8 optics will project disks with radii of 5.42 μm.

At first glance, this may seem to contradict the maxim that resolution is determined by aperture, but we also need to consider how spatial resolution on the sensor translates into *angular* resolution in the sky. Consider the same two instruments we compared at the beginning of Chapter 5, both f/8: a telephoto lens with a 10 mm aperture and 80 mm focal length, and a telescope with 100 mm aperture and 800 mm focal length. We calculated above that both will have Airy disks of 5.42 μm for a point source of light at focus, but the telephoto lens will project a wide swath of sky across the sensor, while the telescope has a much narrower field of view. In each case, stars will appear about 5.42 μm across on the sensor, but stars that are too close together to resolve in the telephoto's image will be more widely separated in the telescope's narrower view. Thus the *angular* resolution is determined by both the focal length, which determines how spatial distances on the sensor correspond with angular distance in the sky, and the focal ratio, which determines the size of the diffraction pattern projected. In order to increase the focal length while maintaining the same focal ratio, the aperture must be increased.

The most commonly used threshold for angular resolution is the Rayleigh Criterion, which considers resolution in the context of two equally bright stars that are very close together. Figure 41 shows a two-dimensional cross section of two Airy disks near each other. If the middle of one star's Airy disk coincides with the first minimum of the other (bottom image), they are just barely distinct as separate points. This is about a 26% drop in light between the maxima. This is the Rayleigh criterion, which is an estimate of the minimum distance between two points where they are clearly discernible as separate. (There are other criteria for minimum resolvable separation, such as Sparrow's limit, but Rayleigh's is more commonly used since it provides a more conservative estimate.)

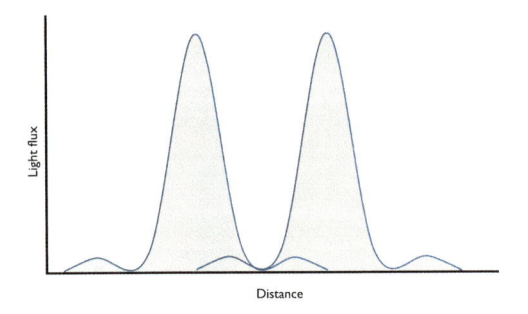

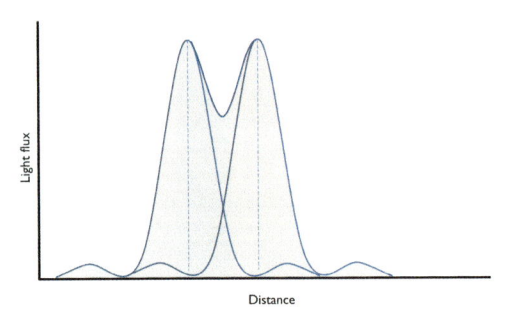

Figure 41. The Rayleigh Criterion

The formula for calculating resolution using Rayleigh's Criterion is

$$Angular\ resolution \cong \sin^{-1}(1.22\frac{\lambda}{a})$$

The result is in radians, so multiply by 206,265 to get the result in arcseconds (57.3 degrees per radian × 3600 arcseconds per degree = 206,265.) The 1.22 is a factor derived from the shape of the curve that marks the second maximum.

Another commonly used formula for resolution, known as Dawes' Limit, uses simpler math and was empirically derived, but it corresponds reasonably well with the Rayleigh Criterion for wavelengths in the middle of the spectrum. Dawes' estimate for the minimum resolvable angular distance in arcseconds (using aperture in mm) is

$$Resolution \cong \frac{115.82}{a}$$

The equations above describe the ideal situation, but we also have the atmosphere interfering with our view. The atmosphere, when clear, is essentially a mildly refractive lens that is constantly varying a little in thickness. This generates a beautiful twinkling of the stars that may inspire poets, but for imagers it means blurring. Places like Mauna Kea and the Atacama Desert of Chile have seeing under one arcsecond, but most of us deal with skies that blur the image of a star over a diameter of one to four arcseconds. At longer focal lengths, the atmosphere becomes the limiting factor.

A two-arcsecond star is one thing in the sky, but when we project it onto a sensor, how big will that image be? Fortunately, there's a simple formula for this too:

$$Size_{projected} = \frac{Size_{angular} \times f}{206265}$$

where angular size is expressed in arcseconds. This formula is a very accurate approximation for angular sizes smaller than a few degrees.

So with a focal length of 1000 mm and average seeing of two arcseconds, a star would form an image about 9.7 μm wide on a sensor at focus. But what is the optimal size of photosites for this telescope? How do we match a sensor to our optics?

Sampling

An important concept from information theory helps us here. The Shannon-Nyquist sampling theorem, which has applications in a wide variety of fields, tells us that if we sample at frequency x, we can resolve objects spaced at least 2x from each other. In other words, we need to sample at a resolution twice that of the smallest resolvable space we wish to reconstruct. There is some debate about applying this theorem to two-dimensional images, with a ratio closer to three times the smallest detail being recommended in some cases as a more practical limit. Here we'll stick with Nyquist's original 2:1 ratio as a conservative threshold for oversampling.

In our case, the atmosphere limits our details to two arcseconds across, thus we want each photosite to correspond with about one arcsecond of sky. The angular space covered by each pixel is called the image scale. We can rearrange the previous equation to get a more convenient form for determining image scale:

$$Image\ scale \cong \frac{206.265 \times pixel\ size}{f}$$

where image scale is expressed in arcseconds per pixel, focal length is in mm, and pixel size is in μm.

Acquiring Images

Given the Shannon-Nyquist theorem, we'd like to have pixels that are half as wide as the seeing. So back to our average two arcsecond skies, we'd want about 1 arcsecond per pixel. This implies that for our 1000 mm scope, we ideally want a pixel size of about 4.85 μm.

If you use a sensor with pixels that are too large for your optics and atmosphere, (or alternately expressed, a scope with too short of a focal length for your sensor), you are not capturing all of the detail you could. This is called undersampling. You lose detail, but by condensing more area of sky into each pixel (decreasing your image scale), but you are getting more signal in each photosite. This improves the signal-to-noise ratio, allowing the faintest regions to better stand apart from the noise. You've made a tradeoff: giving up resolution for an increase in signal. If you go too far, end up with square stars and pixelated details, but this is fine for widefield images. Undersampling is not the worst problem in the world, but you could be doing better in terms of resolution.

If you oversample, on the other hand, you do not gain any additional detail. The atmosphere and optics limit what you can resolve, so a greater resolution of the sensor will show nothing beyond what the properties of atmosphere and diffraction allow. Think of it as a high resolution image of a blurry view: it's still blurry! For daytime photography, this might not be much of a loss, but as we've seen, astronomical images are almost always starved for photons. By oversampling, you are gathering less light in each photosite, meaning your SNR gets lower. At least with undersampling, you were getting something in exchange for your lost resolution (more signal per pixel), but with oversampling, you are losing signal without any gain in resolution. Avoid oversampling whenever possible.

Over the course of a long exposure, the full effects of atmospheric blurring will be visible, even if seeing improves transiently. Most of us have 2–3 arcsecond seeing on an average night, implying appropriate image scales of around 1–1.5 arcseconds per pixel. Those lucky individuals with exceptional atmospheric conditions can use image scales down to around 0.5 arcseconds per pixel.

Figure 42 maps focal length and pixel size to image scale. Most current APS-C sized DSLRs have photosites in the 4–6 μm range, while the full frame DSLRs are typically

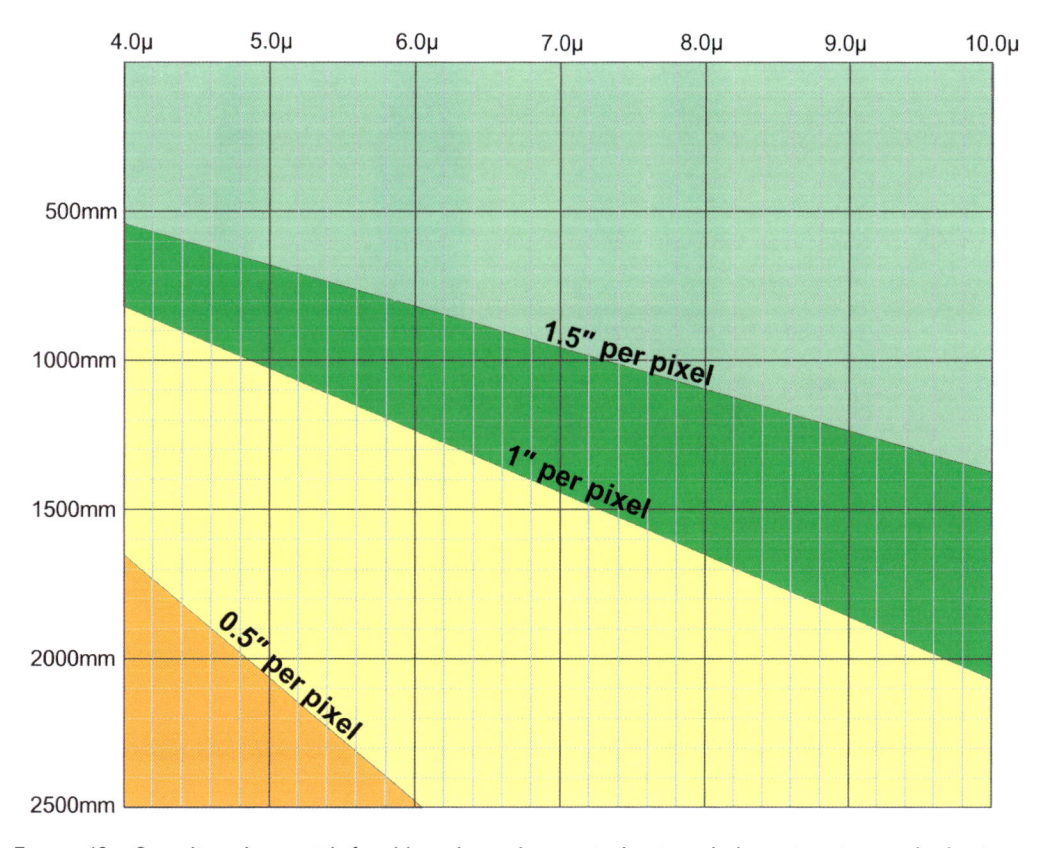

Figure 42. Sampling chart, with focal length on the vertical axis and photosite size on the horizontal axis

larger, up to 9 μm. Dedicated astronomical CCDs vary widely, covering the full range of photosite sizes shown.

> Harry Nyquist's 1928 paper, "Certain topics in telegraph transmission theory," was one of the first to consider the sampling concept, though it was in the context of squeezing a certain number of electrical pulses through a given bandwidth.
>
> Claude Shannon later picked up the problem in 1949 in the context of sampling and reconstruction of a signal. The theorem has uses in many disciplines.

Oversampling revisited

There are three limits to the maximum achievable resolution, but only the largest for a given situation is the rate-limiting step:

1. Movement of air in the atmosphere blurs the view. We measure this in terms of angular resolution. For most places on earth, it is in the low single-digit arcseconds.
2. The optics limit resolution in two ways. First, the aperture determines the maximum angular resolving power. But the focal length determines the projected size of that resolution on the sensor.
3. Finally, the size of each photosite on the sensor determines how the projected image is sampled in terms of spatial resolution.

Imagine the sky as a giant monitor with a fixed resolution or pixel size, as in Figure 43. Of course real optical resolution does not manifest itself with hard edges like a grid of pixels, but the basic principle holds. Atmospheric seeing conditions dictate the resolution of this giant monitor. Let's say you have skies with two arcsecond seeing. In other words, a point source of light is spread into a blur two arcseconds across. No matter how hard you or you or your sensor look, this is the ultimate limit on what you can see. (There are techniques like "lucky imaging" that take advantage of brief moments of great seeing for short exposures, but without adaptive optics, this is the best you can do for long exposure images.)

Diffraction means that your optics also impose a limit on the resolving power, which we've seen is determined by the aperture and the wavelength of light being refracted. For this example, let's assume you use a 100 mm refractor. Only when the seeing is excellent and the optics are small do optics ever become the rate limiting step, however. Any telescope optics greater than about 65 mm in diameter will have limiting angular resolution smaller than two arcseconds across the visible spectrum. There is no SNR "penalty" for having optics with a theoretical angular resolution greater than the seeing—oversampling applies only to sensors.

Using the Nyquist sampling theorem as a guide, the sensor shouldn't have photosites much smaller than half the width of the resolution in the view being provided. Since the optics are not limiting resolution in this case, we should use a sensor with photosites no smaller than a projected size of one arcsecond for this telescope's focal length. If the photosites are any smaller, no additional resolution is gained,

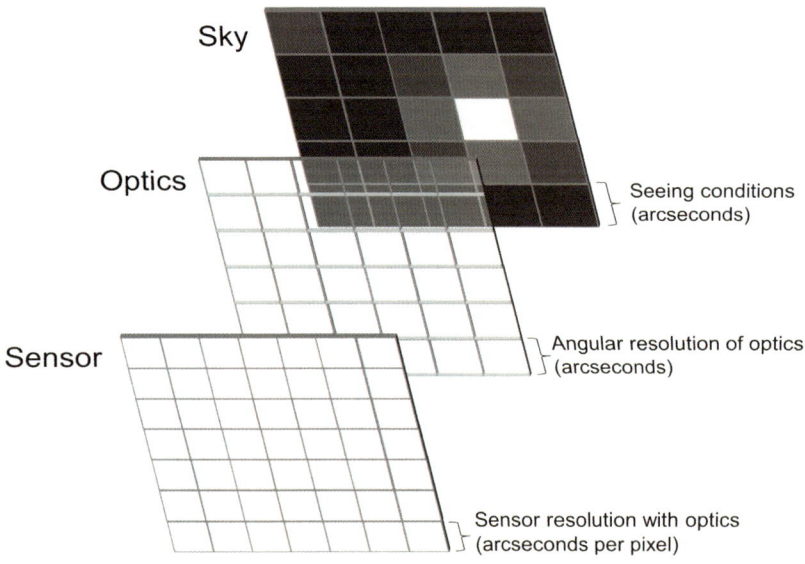

Figure 43. The three possible limiting resolutions in imaging

but fewer photons are collected at each site, which means reduced SNR, degrading the final image. Some believe that the Nyquist theorem is too conservative for practical use. Or perhaps you would rather use Sparrow's Limit than the Rayleigh criterion for resolution. Either way, the results won't change dramatically. The Shannon-Nyquist theorem gives us a good rule of thumb: the photosites should not be substantially smaller than half the maximum resolution of the projected image, whether that is limited by the optics or (more likely) the atmosphere.

The atmosphere is the ultimate resolution limit, which is why we have telescopes in space and adaptive optics on Earth. For aesthetic imaging, undersampling is generally okay; finer details do not necessarily create better images. It's usually more important to improve the signal-to-noise ratio. But using a sensor with a resolution greater than the atmosphere or optics will allow—oversampling—is a problem. No matter how small the photosites are, no smaller detail can be captured than what the atmosphere and optics allow. So no true increase in resolution is gained, yet the loss in SNR remains.

Exercises

2.1 Using a refractor with a 100 mm diameter objective under skies with two-arcsecond seeing, what is the smallest approximate photosite size if the refractor is f/5 and you want to sample no more than twice the seeing? What if it is f/10?

2.2 Assuming that we are imaging in visible light from 400–700 nm in wavelength, what is the smallest diameter aperture that will have better than one arcsecond resolution across the visible spectrum?

Field of view

Choosing a sensor is about more than pixel size, it's also about the size of the whole sensor. Focal length and sensor size determine the field of view. Long focal length telescopes not only require larger photosites in order to be properly sampled, but having a larger chip is also helpful to

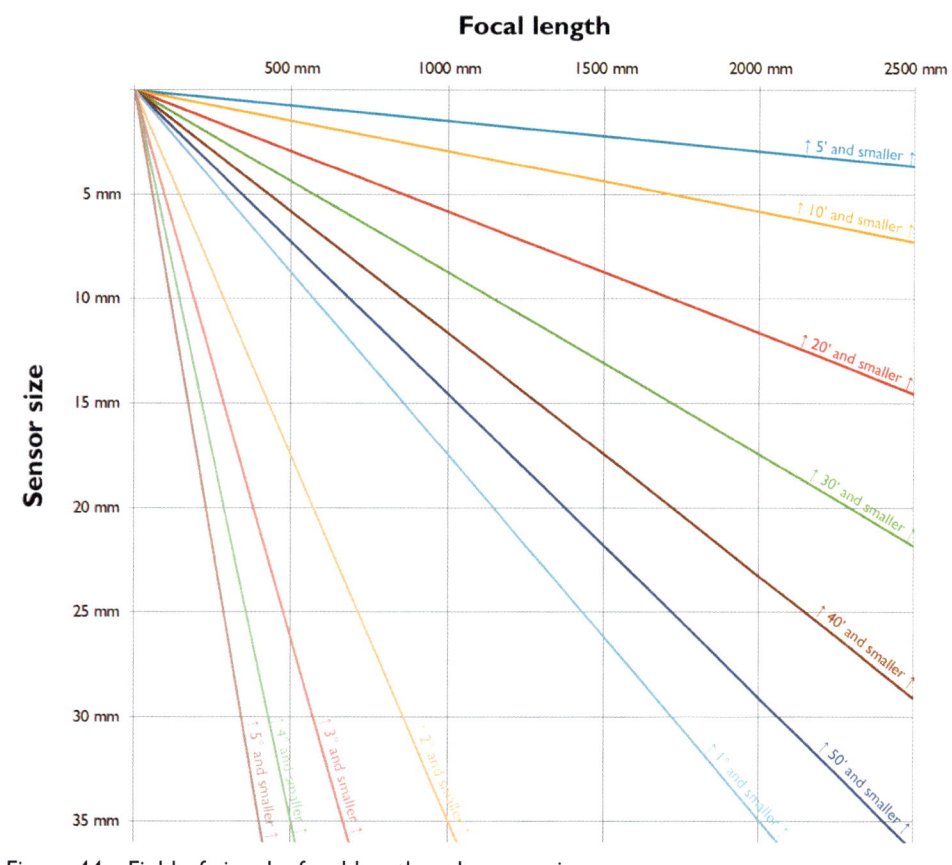

Figure 44. Field of view by focal length and sensor size

Focal Length	Kodak KAI-2020 11.8×8.9 mm	Kodak KAF-8300 18×13.5 mm	Canon APS-C 22.3×14.9 mm	Nikon/Pentax APS-C 23.7×15.7 mm	Canon APS-H 27.9×18.6 mm	Full Frame 36×24 mm
200 mm	3.38°×2.55°	5.15°×3.87°	6.38°×4.27°	6.78°×4.5°	7.98°×5.33°	10.29°×6.87°
300 mm	2.25°×1.7°	3.44°×2.58°	4.26°×2.85°	4.52°×3°	5.33°×3.55°	6.87°×4.58°
400 mm	1.69°×1.27°	2.58°×1.93°	3.19°×2.13°	3.39°×2.25°	4°×2.66°	5.15°×3.44°
500 mm	1.35°×1.02°	2.06°×1.55°	2.56°×1.71°	2.72°×1.8°	3.2°×2.13°	4.12°×2.75°
600 mm	1.13°×0.85°	1.72°×1.29°	2.13°×1.42°	2.26°×1.5°	2.66°×1.78°	3.44°×2.29°
700 mm	0.97°×0.73°	1.47°×1.11°	1.83°×1.22°	1.94°×1.29°	2.28°×1.52°	2.95°×1.96°
800 mm	0.85°×0.64°	1.29°×0.97°	1.6°×1.07°	1.7°×1.12°	2°×1.33°	2.58°×1.72°
900 mm	0.75°×0.57°	1.15°×0.86°	1.42°×0.95°	1.51°×1°	1.78°×1.18°	2.29°×1.53°
1000 mm	0.68°×0.51°	1.03°×0.77°	1.28°×0.85°	1.36°×0.9°	1.6°×1.07°	2.06°×1.38°
1200 mm	0.56°×0.42°	0.86°×0.64°	1.06°×0.71°	1.13°×0.75°	1.33°×0.89°	1.72°×1.15°
1400 mm	0.48°×0.36°	0.74°×0.55°	0.91°×0.61°	0.97°×0.64°	1.14°×0.76°	1.47°×0.98°
1600 mm	0.42°×0.32°	0.64°×0.48°	0.8°×0.53°	0.85°×0.56°	1°×0.67°	1.29°×0.86°
1800 mm	0.38°×0.28°	0.57°×0.43°	0.71°×0.47°	0.75°×0.5°	0.89°×0.59°	1.15°×0.76°
2000 mm	0.34°×0.25°	0.52°×0.39°	0.64°×0.43°	0.68°×0.45°	0.8°×0.53°	1.03°×0.69°

accommodate the narrow field of view. The precise formula for field of view in degrees is

$$FOV_{angular} = 114.6 \tan^{-1} \frac{d}{2f}$$

where d is the size of one dimension of the sensor, and f is the focal length of the instrument. Despite the trigonometric term however, the relationship is very nearly linear for all but the shortest focal lengths. The simpler approximation

$$FOV_{angular} \cong 57.3 \frac{d}{f}$$

is accurate to within 0.3% for focal lengths over 200 mm for all sensors up to full frame (36 mm). It only differs meaningfully from the precise formula for very short focal lengths, so use the precise formula for wide angle lens calculations.

Figure 44 graphically illustrates the relationship between focal length, sensor size, and field of view. Typically the longer axis of the sensor would be used to determine whether an object of a given size will fit. In order to frame an object and allow for some alignment margin, you'll need a slightly shorter focal length for an object that spans the whole sensor.

The accompanying table shows the field of view for common sensors at various focal lengths. DSLRs come in only a few basic sizes: full-frame (Canon 5D, etc), a few variations on APS, and Four-Thirds. Dedicated astronomical CCDs come in an endless range of sizes, but two common sensors are shown. The Kodak KAF-8300 is a popular astronomical sensor, equivalent in size to a Four-Thirds DSLR sensor, found in cameras like Orion's Parsec, QSI's 583 series, and SBIG's ST-8300 series. The KAI-2020 is a much smaller, but still commonly used sensor, found in the popular SBIG ST-2000 series. Cameras using the Kodak KAI-11000 series CCDs are approximately full-frame in size.

Equipment recommendations

All of this leads to some insights about the best equipment for an imaging setup. First, spend your money on a quality mount—a cheap mount will lead to needless frustration. A stable mount is central to obtaining good quality images. All the processing in the world can't make a great final image from poorly-tracked exposures. If you have an initial pool of money for starting out, spend half of it on the mount, and split the remainder on the telescope and camera.

After buying the best mount you can, it should be clear by now that short focal lengths and large pixels are the low stress approach to imaging. Short focal lengths are forgiv-

Acquiring Images

| Equipment Check | Focal reducers and field flatteners |

Since focal reducers are designed to work within specific optical parameters, there are no truly universal options. Many, however, are designed to work with a range of telescopes. It can be a matter of trial and error to find a match and also to determine the optimal distance from the sensor. The best focal reducers are designed to work with a particular telescope and with specific spacing.

For years SCTs have had reducers of varying quality that screwed directly onto the back. These typically reduce an f/10 telescope to either f/6.3 or f/3.3, with limited success at the latter ratio. (A better way to get super-fast performance from an SCT is to use the Hyperstar or Fastar system.) Newer SCTs like Celestron's EdgeHD already correct for field curvature, so a dedicated reducer is required. Most Newtonians have naturally fast optics, so there are few focal reducers designed specifically for them. The best Newtonian reducers for imaging incorporate Wynne corrector optics with the reducer, as with those from AstroSysteme Austria (ASA). While expensive, these can be used to create extremely fast astrographs like the Boren-Simon PowerNewts.

Refractors have many more options, most powered for a 0.65–0.85× reduction. They are usually designed to work within either a range of focal lengths or a range of focal ratios, but some are designed for a specific scope. William Optics has offered four versions of their refractor reducer-flattener over the years, with the most recent edition including an adjustable backfocus distance. Televue also makes several high quality reducer-flatteners. The TRF-2008 was originally designed for their smaller TV-76 and TV-85 refractors, but imagers have found that it works well with a wide range of apochromatic refractors. Televue lists it as compatible with scopes in the 400–600 mm range. For those in the 800–1000 mm range, they offer the RFL-4087. Because of their flat field designs, the NP-series scopes from Televue have their own dedicated reducer, the NPR-1073. Takahashi makes a variety of reducers, but they are designed with specific Takahashi telescopes in mind. Borg makes a triplet 0.7× reducer (model 7870) that allows adjustment to specific focal lengths between 325 mm and 1000 mm to better match the telescope. Astro-Physics makes the 0.67× CCDT67 reducer, which has been found to work well with Ritchey-Chrétiens, as well as the 0.75× 27TVPH for larger sensors and several other reducer-flatteners designed for specific models.

Two popular non-reducing field flatteners are the AstroTech AT2FF and the HoTech SCA. While both were designed to work with refractors, they have also been found to be successful with other designs like the Ritchey-Chrétien. The AT2FF is specified to work within the f/6 to f/8 range, and the SCA claims an f/5 to f/8 range.

ing of tracking error, seeing is less of a factor, and many of the most beautiful targets in the sky are very large. Start with optics under 750 mm. Larger pixels (or binning pixels 2×2) can also help if the optics aren't as fast as you'd like. From a budget perspective, the best values lie where there are economies of scale. There are dozens of 80 mm f/6 refractors available, many of exceptional quality, that can be had at a bargain. Paired with a DSLR, this probably represents the best entry-level value in imaging. One could spend years exhausting the list of targets suitable for this combination.

When you are ready to capture the details of smaller galaxies and planetary nebulae, longer telescopes and larger mounts are required, but to begin with them is likely to bring needless frustration.

Focal reducers and field flatteners

Focal reducers optically shorten the focal length of a telescope, yielding a 'faster' effective focal ratio. Some focal reducers also act as field flatteners, bringing the edges of the field of view closer to the focus plane of the center. There are also dedicated field flatteners that do not alter the focal ratio.

Only a few telescope designs produce a flat field without an additional flattener close to the point of focus. The best reducers and flatteners are designed to precisely match a specific telescope or are incorporated into the system's design. This is the case with the modified Petzval designs used by high-end astrographs like the Takahashi FSQ series and Televue's NP series. Refractors with only objective lenses inherently suffer from field curvature, so they need a flattening lens at minimum. (The radius of the curvature for most refractors is about one-third the focal length.)

Most standalone reducers are designed to work within a range of focal lengths or ratios. In order to get the best performance out of any reducer or flattener, it is crucial to ensure that the specified reducer-to-sensor distance is achieved. Incorrect spacing can lead to coma and field curvature, and the spacing needs to be more precise the faster the optics are. The flattening element is most sensitive to spacing. Reducers that are not designed to flatten the field are less sensitive to spacing, though the degree of reduction will vary with the distance from the sensor.

Figure 45 shows a simplified schematic of how a focal reducer works in a triplet refractor. A typical reducer-flattener consists of a pair of lenses, one negative, one positive, but some designs use three or even four lenses. The focal

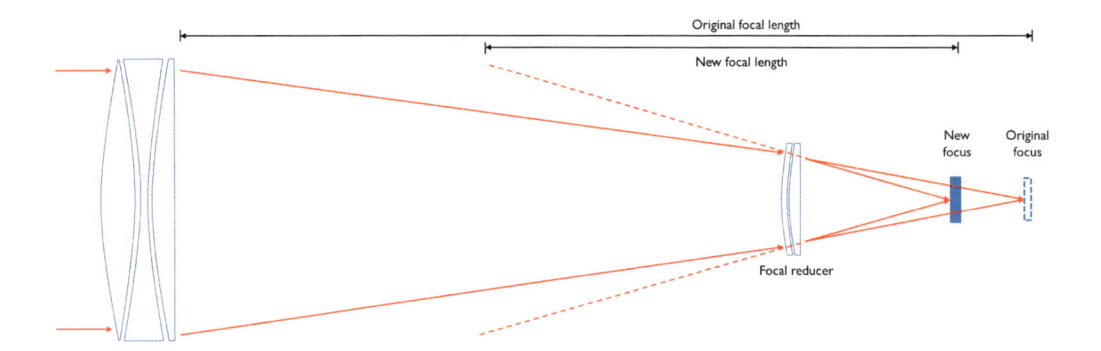

Figure 45. Schematic of a focal reducer's effect

ratio determines how steeply the light cone converges on the sensor. By shortening this cone and bringing the rays to focus at a closer point with a focal reducer, the resulting projected image is concentrated as if it had come from a shorter focal length but the same aperture. (The dotted lines show the implied focal length with the focal reducer in place.)

Considering that no additional photons entered the telescope, it would seem that a reducer can't actually brighten the image, only concentrate the same light into a smaller area of the sensor. So where's the free lunch? It's in the light that falls around the sensor. In most telescopes the image circle at focus is usually larger than the diagonal of an APS sized sensor. If your sensor is only capturing a small portion of the light cone, then a focal reducer can indeed put more photons onto the sensor, providing a wider field of view and concentrating more photons into each photosite. It can focus the light that would normally fall outside of your sensor and project it into a tighter cone—more light is actually falling on the sensor because less is being wasted.

Figure 46 shows the same area of sky around Deneb at equal exposure times through a small refractor at f/6.3 without a focal reducer and at f/5 with a reducer. Careful inspection will reveal the wider field of view (and thus larger image scale), but it is immediately obvious that the sky background is brighter at f/5. The inset histograms quantify this, with the mean brightness on an 8-bit scale growing from approximately 52 to 68.

This is not true for every system or every sensor. If the sensor is as large as the usable image circle, a focal reducer will cause vignetting. The size of the usable image circle for an optical system is not a straightforward value to calculate, and some vignetting is correctable in post-processing, so

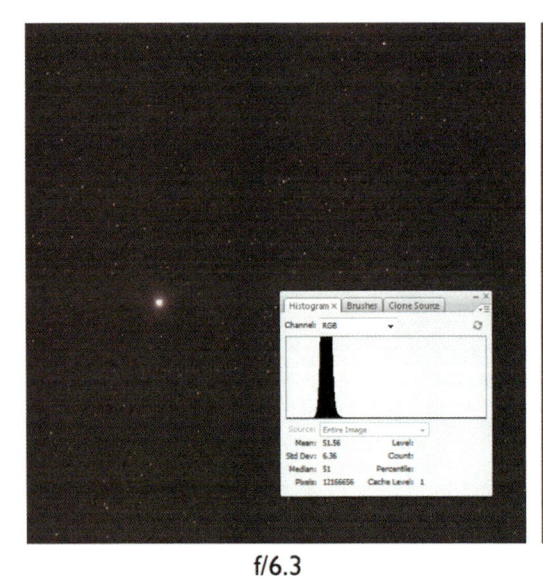

f/6.3   f/5.0

Figure 46. Focal reducer comparison

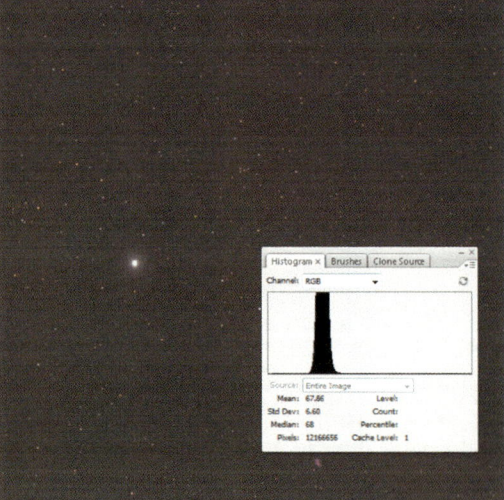

it is hard to determine whether a given reducer will work with a telescope and sensor. In general, reducers greater than about 0.7× are not practical for most systems unless the sensor is small. A telescope's normal projected image size is shrunk by the reducer's power. For example, a telescope that normally produces a 40 mm usable image circle will produce a 32 mm circle with a 0.8× reducer. The distance between the reducer and the sensor also influences the reducing power, so there is some variability depending on where you place it.

So why do we sometimes need focal reducers? Why not just make the primary mirrors or objective lenses faster? It is far more expensive to make large lenses and mirrors with lower f-ratios. Fast optics have more curvature than slower ones, and it is challenging to manufacture primary optics with precise curvature while suppressing ratio-dependent aberrations like field curvature, chromatic aberration, coma, and astigmatism. To achieve a lower focal ratio for most systems, it is far easier to add a focal reducer near the sensor.

> Planetary imagers make the opposite tradeoff as deep-sky imagers. Because of the small size and high surface brightness of planets, imagers often capture hundreds of short-exposure images at very high focal lengths, often imaging at f/40 or greater. To do this, they use the opposite of a focal reducer: a Barlow lens.

7 Choosing appropriate objects to image

Getting a sense of scale

Looking through an astronomy book or magazine, it is easy to completely lose any sense of scale. Observatory images show tiny planetary nebulae enlarged to show detail that could never be revealed by a backyard telescope. What is termed "wide field" for the Hubble Space Telescope would be a very narrow field for most amateur optics. It can be difficult to tell which objects will yield a richly detailed image and which will show up as an indistinct speck. To create great images, it is important to choose targets that are the right size for your optics and sensor.

Figure 47 shows some commonly imaged objects at the same image scale, with the moon (about 30') in the center for comparison. Note how small the two frequently-imaged planetary nebulae, M57 and M27, are. M31 and M33 are the two biggest galaxies in the northern sky; most galaxies are much smaller than M51, requiring long focal lengths to resolve any detail.

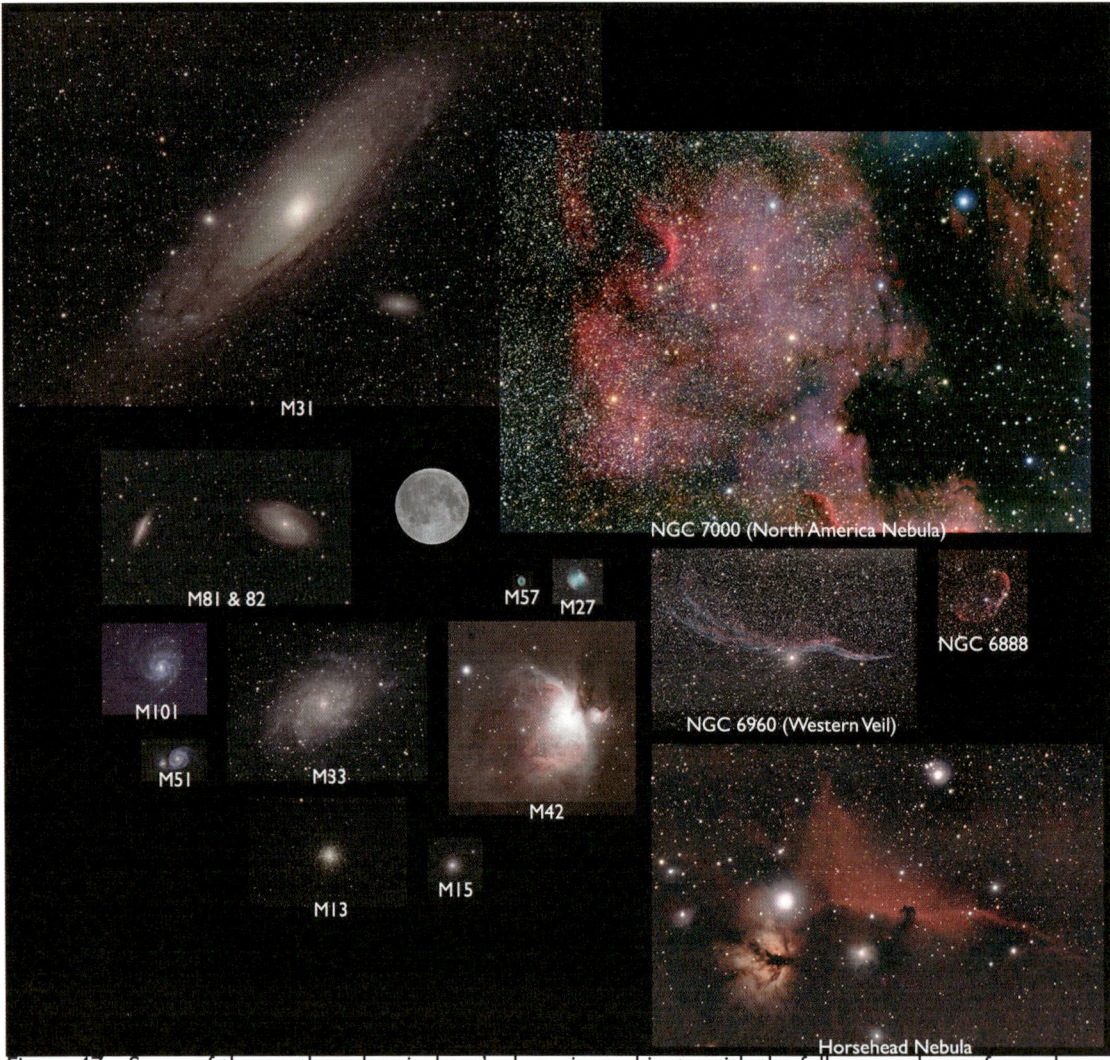

Figure 47. Some of the northern hemisphere's showpiece objects, with the full moon shown for scale

Acquiring Images

Catalog	Description
Arp	In 1966 Halton Arp published *Atlas of Peculiar Galaxies*, containing images of 338 galaxies that fell outside of the standard elliptical or spiral shapes. Many of these are interacting or colliding galaxies. For those with telescopes large enough to resolve them well, some of the Arp galaxies are beautiful targets.
B	E. E. Barnard cataloged 370 dark nebula in a 1919 publication. These clouds of interstellar dust block the light of stars and nebular emissions behind them from our perspective. They are generally in the plane of the Milky Way. Some are enormous, and these subtle objects can make excellent, though challenging, imaging targets.
Caldwell	Patrick Moore (actually, Caldwell-Moore, hence the catalog name) compiled a list of 109 bright deep sky objects that Messier missed. Published in 1995, it is a hit list for amateur visual observers that includes southern hemisphere objects and northern hemisphere objects that Messier didn't include.
Ced	Sven Cederblad's 1946 "Catalog of Bright Diffuse Galactic Nebulae" lists 215 nebulae. Most Ced objects are more commonly referred to by another catalog designation.
LBN	A rarely used general catalog for nebulae from B.T. Lynd's 1965 publication "Catalogue of Bright Nebulae."
Messier	The oldest of the catalogs still in use, Charles Messier's list was originally designed to be a list of fuzzy objects that could be mistaken for a comet, to aid in their search. Messier's original list had 103 objects, with M104-M110 added by others. Because the objects were found with long focal length refractors in the 18th century, it's a mixed bag from an imaging perspective. Some are stunning (M31), while others are dubious (M40 is a double star).
RCW	Rodgers, Campbell, and Whiteoak published a catalog of southern hemisphere emission nebulae in 1960. Like the Sh2 catalog with which it has some overlap, it is a great source for narrowband imaging targets.
UGC	The Uppsala General Catalogue of Galaxies (UGC) lists 12,921 northern hemisphere galaxies. For most imagers, these are the designations listed for tiny background galaxies.
vdB	Sidney van den Bergh's (vdB) catalog, originally published in 1966, lists 159 reflection nebulae visible from the northern hemisphere.
Sh2	The second edition of Stewart Sharpless's (Sh2) catalog lists 313 emission nebulae visible from the northern hemisphere. It was published in 1959, and it contains a trove of beautiful narrowband objects.

The deep-sky catalogs

The most commonly used catalogs of astronomical objects are probably the *New General Catalogue* (NGC) and *Index Catalogues* (IC), both created under the leadership of John Louis Emil Dreyer over 100 years ago. The NGC was compiled in 1888, before photography was applied to astronomy, so it is a list of visually discovered objects. This limitation excludes many objects with low surface brightness, especially large emission nebulae. Nonetheless, it was an amazing accomplishment for the time, considering that Charles Messier published his catalog of just over 100 objects a century before, and Dreyer's catalog listed 7840.

Dreyer later added more visual objects with the *Index Catalogue* in 1895, and even more with the *Second Index Catalogue* in 1908. The second IC is noteworthy because it was the first catalog to incorporate photographically-discovered objects, hence many of the finest emission nebulae are designated with IC numbers.. These two catalogs are now simply known as one, containing 5,386 objects.

There were of course errors and overlaps between the catalogs, and revisions and corrections were made throughout the 20th century by many authors. The most comprehensive and recent revisions have come from Dr. Wolfgang Steinicke with the *Revised New General Catalogue and Index Catalogue*, which is the basis for many current software object databases. Even with over a century of revisions, about 2% of the entries are simply lost or not yet found based on erroneous original location data or observer error.

In addition to the NGC and IC catalogs, other catalogs have been created, usually for specific object types, which may have some overlap with entries in the NGC/IC. There are dozens of other catalogs, usually of a specific object type and taken from a specific photographic survey. Some of the most commonly used are listed in the table.

> A difference of one magnitude visually represents a brightness difference of 2.51 times. Why 2.51? The ancient Greek practice of dividing the stars into six groups based on brightness was formalized in 1856 by the British astronomer Sir Norman R. Pogson. He set sixth magnitude stars at exactly 100 times the brightness of those of first magnitude. Since a change of five magnitudes meant a factor of 100, a change in a single magnitude was therefore the fifth root of 100, or 2.51.

A survey of object sizes

In order to get a sense of how many deep-sky objects there are at various sizes, we'll use Dr. Steinicke's *Revised New General Catalogue and Index Catalogue* as the broadest current and accurate catalog of deep-sky objects. There are 14,002 entries in the 2012 update. Nearly every object worth observing is listed, providing a good sample from which to evaluate the size ranges of deep sky objects. It is important to note that groupings of objects are not accounted for here—these are individual object sizes. For example, what we refer to as the Rosette Nebula actually covers several NGC/IC entries. Where this is the case for a commonly imaged object, that is described in the sections below.

Since we are discussing imaging targets, many loose open clusters or asterisms without associated nebulosity were left out, as were some questionable nebulae in the IC, all "lost" entries, and entries without a known size.

Giant Objects (2° or greater)

At this scale, telephoto lenses and short focus refractors are generally needed. Most objects this large are deep-sky highlights, and because of their size, great detail can be revealed.

There are two galaxies greater than two degrees across in the NGC/IC. These are the Small Magellanic Cloud (NGC 292) at about five degrees across and the Andromeda Galaxy (NGC 224, M31) at about three degrees. An obvious omission from the NGC/IC is the Large Magellanic Cloud at about 10° across. The Magellanic Clouds are telephoto lens targets, while the Andromeda Galaxy can just barely be squeezed onto an APS-sized sensor at 500 mm focal length.

There are also several large nebulae in this range, and all are beautiful imaging targets. The largest is the very faint Witch Head Nebula (NGC 1909, IC 2118), followed by the California Nebula (NGC 1499), the Eagle Nebula (NGC 6611, M16), the Eta Carina Nebula (NGC 3372), and the North America Nebula (NGC 7000). IC 1396 is a huge emission nebula complex in Cepheus that spans about three degrees and contains the commonly imaged Elephant's Trunk formation. IC 4592 is an infrequently imaged reflection nebula that spans more than two degrees in Scorpius. It is adjacent to several dark nebulae and another large reflection nebula, IC 4601, making a stunning telephoto target.

Not included in the NGC/IC are many other beautiful objects and grouping of objects that extend more than two degrees. To name a few in the northern hemisphere: The Pleiades (M45) are nearly two degrees across, and the surrounding dark nebula span even further. The Orion Nebula (M42) is only about 65', but capturing the larger complex with M43 and other nebulosity extends over 120'. Widefield shots of the Virgo cluster of galaxies, including Markarian's Chain, can extend this far as well. Even though each individual component of the Veil Nebula is less than a degree, the whole complex in Cygnus spans several degrees. And the pentagonal Rho Ophiuchus star formation complex is a sampler of deep-sky objects in a five-by-five degree area, including several nebulae, dust lanes, and even the globular cluster M4.

	Revised NGC/IC Objects			
	Galaxies	Star Clusters	Diffuse Nebulae	Planetary Nebulae
2°+	2	-	10	-
1–2°	1	7	11	-
30–60'	1	25	27	-
15–30'	21	102	29	1
5–15'	251	263	56	3
1–5'	6678	91	69	35
<1'	4373	4	8	99

Acquiring Images

Catalogs are not always completely trustworthy as sources for imaging targets. The IC lists several large and apparently very diffuse nebulae that do not appear as discrete objects. Several of these are in Taurus (IC 341, IC 353, IC 354, and IC 360). IC 353 appears to be an extension of the Pleiades reflection nebulosity, but the identity of the others is less clear. IC 1831 in Cassiopeia is another example of these 'barely there' objects. It seems to be an extended area of the much brighter Heart Nebula, which is oddly not part of the catalog (appearing instead in the Sharpless catalog).

Really Big Objects (1–2°)

There are several more deep sky highlights in this range, and these are generally captured well by short refractors. The sole galaxy in this group is the Triangulum Galaxy (NGC 598, M33) at about 70'. While large, its surface brightness is lower than M31, so it can be a more difficult target.

The rest of the NGC/IC objects of this size are nebulae, and again, these are some of the most beautiful and most frequently imaged objects in the sky. NGC 7822 in Cepheus is an emission nebula with complex pillars of gas. With Sh2-171 (a.k.a. Ced 214) and Sh2-170, a widefield image of the area oriented properly looks like a punctuation mark dubbed the "Cosmic Question Mark." The huge Rosette Nebula in Monoceros accounts for five NGC numbers and spans almost a degree and a half. Even though its components are listed individually under 50', the Lagoon Nebula (M8) in Sagittarius (NGC 6523 and 6526 with the cluster 6530) spans over 90' when they are considered together. The Orion Nebula (NGC 1976, M42) is quite possibly the most imaged deep-sky object. The first deep-sky photograph ever taken, by Henry Draper in 1880, was of M42.

Also noteworthy are the individual components of the Veil Nebula, NGC 6960 and 6992, which are photogenic on their own. Also in Cygnus, near Deneb and the North America Nebula, is The Pelican Nebula (IC 5070 and 5067). The Horsehead Nebula (IC 434) should be part of everyone's personal image collection, producing equally interesting close-ups images and widefield shots that include the nearby Flame Nebula (NGC 2024).

To the south, emission nebula IC 4628, the "Prawn Nebula," lies in the tail of Scorpius and spans a degree and a half. Known as the "Southern Pleiades," IC 2602 is a large open cluster in Carina. Corona Australis features the reflection nebula NGC 6727, which is part of a larger complex with NGC 6726 and several dark nebulae.

Objects from 30-60'

The sole NGC/IC galaxy in this size range is the bright, irregular Sculptor Galaxy (NGC 55), but there are many more nebulae worth imaging here, including the nebulosity associated with the Pleiades, NGC 1435, the Merope Nebula or Tempel's Nebula. There are also about two dozen modestly interesting open clusters over half a degree across.

The Heart (Sh2-190)and Soul (Sh2-199) Nebulae in Cassiopeia are two of the finest narrowband objects in the sky, and both are just under a degree in size. Neither are technically in the NGC or IC, but parts of them have designations in these catalogs. (The brightest part of The Heart is NGC 896, and IC 1848 is the cluster within The Soul.) To capture them together requires more than a three degree field of view.

The Tarantula Nebula (NGC 2070) is a very bright emission nebula in the Large Magellanic Cloud that is also known as 30 Doradus, a designation from when it was originally considered a star. The two largest globular clusters are also southern delights. Omega Centauri (NGC 5139) and 47 Tucanae (NGC 104) are each almost a full degree across, about twice the size of the full moon.

Objects from 15-30'

Now that we are getting into the smaller objects, there are many more, and this range is something of a sweet spot for medium scopes (5–8 inches) paired with large sensors. The largest planetary nebulae, the Helix Nebula (NGC 7293) falls into this category at 17' across.

Two of the best Messier galaxies, The Pinwheel Galaxy (M101, NGC 5457) and Bode's Galaxy (M81, NGC 3031), are just under 30', and both lie in Ursa Major. Twenty other galaxies lie in this size range, nearly all of which are worth imaging. There are 29 globular clusters, including the fine M13 in Hercules, and over seventy open clusters. There are also 29 emission nebulae to choose from in this range.

Objects from 5-15'

There are 251 galaxies between 5' and 15', including most of those in the Messier catalog. The Whirlpool Galaxy (M51, NGC 5194) and the Sunflower Galaxy (M63, NGC 5055) are just two of the many highlights.

There are a few planetary nebulae here, including the much-imaged Dumbell Nebula (M27, NGC 6853) and the more obscure Fornax Nebula (NGC 1360), both around seven arcminutes across. The Crab Nebula (M1, NGC 1952), with its complex filamentary structure, falls here as well.

| Equipment Check | Books and software for choosing a target |

There are many approaches to choosing a target, but the products available fall into two categories: software and books.

Planetarium software can be very helpful for finding objects and planning an imaging run. The options here are limitless. There are feature-packed retail packages that integrate with image capture software, like *TheSky* and *Starry Night*, but there are also many free programs available like *Stellarium* and *Cartes du Ciel* that rival the commercial programs with their features.

Any planetarium software can help you find an object's position on a given night, but there are also programs designed specifically for astrophotography. A few examples to look into are *Deep-Sky Planner* from knightware, *AstroPlanner* from iLanga, and *Sky Tools* from Skyhound. These can graphically illustrate every aspect of an object that will affect imaging, highlighting the best opportunities for capturing a good image.

The old manual method of looking in a book also works, and there are so many options here, you could literally fill a library with them. Unfortunately, most published atlases focus on visual objects, overlooking some of the most beautiful imaging targets in the sky. There is little reason for most visual observers to seek out some of the large emission nebulae or dark nebulae, yet these can be showpieces for imagers.

There are 263 star clusters, including at least 70 globulars, and 56 emission and reflection nebulae. It's important to bear in mind how small some of these objects are, especially when looking at high resolution images captured by very long focal length telescopes. For instance, the frequently imaged Sombrero Galaxy (M104, NGC 4594) is only about eight arcminutes across, and the Crab Nebula is only six arcminutes. That said, there is an abundance of choice as you get smaller in scale (and hopefully larger in optics).

Objects from 1-5'

More than half of the NGC/IC falls into this range, the vast majority of which are galaxies, with over 6600 to choose from. Unfortunately, these are small objects that require very large apertures to resolve. Despite their size, there are still highlights worth seeking out, especially planetary nebulae, of which there are 35. Consider that the Ring Nebula (M57, NGC 6720), one of the finest objects in the sky, is only 1.4' across. When resolution is a problem, alternate methods may be needed to reveal details, especially drizzle processing. Stable, transparent skies are also crucial, so imaging close to the zenith can help. For the brightest of these small objects, "lucky imaging" is another more advanced technique worth looking into.

Objects smaller than 1'

While this list is theoretically almost infinite, the NGC/IC lists about 4484 objects, nearly all of which are small galaxies. A very large telescope is required to capture any detail on objects this small, and even then, atmospheric seeing limits capturing much detail.

In spring and autumn, our view of the night sky looks away from the disk of the Milky Way, whose dust obscures the view beyond our galaxy. Spring in particular is known as "galaxy season" in the northern hemisphere, as it brings the Virgo and Coma galaxy clusters into view.

As we look toward the center of the Milky Way in the summer months of the northern hemisphere, many globular clusters and planetary nebulae come into view.

Emission nebulae are also residents of our galaxy, though more evenly distributed, so both summer and winter offer plenty of to choose from.

8 FOCUSING AND AUTOGUIDING

Focusing

Astronomical imaging requires a very high level of focus accuracy. Even the slightest softness in focus can noticeably reduce the image quality, and this can occur when the focus is only a fraction of a millimeter off. Focus sensitivity increases exponentially as the focal ratio decreases—a 0.1 mm shift in focus affects the resolution of an f/6 system four times as much as an f/12 system. This is illustrated by Figure 48. On the top row is the hypothetical view of a star through an f/6 refractor at perfect focus, at 0.1 mm out of focus, 0.2 mm out of focus, and 0.3 mm out of focus. For comparison, on the bottom row is an f/12 refractor of the same aperture at the same distances from focus. The fourfold difference in sensitivity to focus is clearly visible.

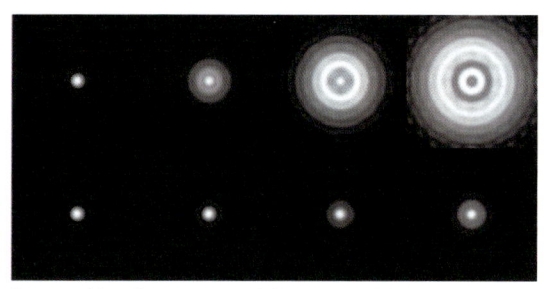

Figure 48. Stars out of focus by 0.0. 0.1, 0.2, and 0.3mm at f/6 vs f/12 (views simulated)

This sensitivity of focus to focal ratio makes sense considering that the focal ratio determines the angle of convergence for the light cone (Figure 49). Any point where the light cone is narrower than the diameter of the Airy disk is generally considered to be in focus. Since the size of the Airy disk is also dependent on the focal ratio (smaller for shorter ratios), there are two factors driving a narrow range of accurate focus for short focal length systems.

Because of this sensitivity to focus, even thermal expansion and contraction can cause focus to shift over the course of a night of imaging. A telescope that is focused perfectly at the start of an imaging session can drift out of focus if the temperature changes or the scope was not equilibrated before initial focusing. Telescopes have different sensitivities to temperature depending on their tube material, design, and optics, but even the least sensitive should only be focused after it has equilibrated to the ambient temperature.

Figure 50 shows crops from unprocessed four-minute exposures of M78, a reflection nebula in Orion, taken at 40-minute intervals. The refractor was not well-equilibrated to the outside temperature before focusing, so while focus was sharp for the first exposures, it drifted as the evening got colder. In the first image, even a single subexposure gives a pretty good impression of the nebula's structure, and several small stars are noted with arrows. Forty minutes later, the nebulosity has lost its shape, and those dim stars are almost gone. By 80 minutes, we can't make a fair comparison, because the target moved into a more light polluted

Figure 49. Focus tolerance at f/2 and f/8

area of the sky, b
but a vague fuzz
no detail availabl
star in the lower

t=0　　　t=40 min　　　t=80 min　　　t=120 min　　　t=160 min

Figure 50. Focus shift from thermal contraction over time

Achieving critical focus

Given its importance, there are many tools available for achieving accurate focus. For cameras linked to a computer, any of several software packages can monitor the width of stellar images and signal when the best focus is achieved. When connected to a motorized focuser, the process can be automated. An alternative approach is to use a focusing mask, which requires some human intervention, but is a cheaper approach that works just as well.

Motorized focusers can be very expensive, but they provide the ability to automatically and precisely focus. There are two types of motors available: stepper motors and servo motors. Steppers are more expensive than servos, but they can be directed to repeatable positions. This makes them easier to automate than servos, so if you plan to use a computer to drive your focuser, steppers are generally worth the additional cost.

To control a motorized focuser, the most widely-used focusing application is FocusMax. FocusMax is a free application written by Larry Weber and Steve Brady that uses ASCOM drivers to interface with imaging software and a motorized focuser to automatically determine and set optimal focus. The software first iteratively characterizes the optical system to a create "V-curve" that compares the focuser position on the x-axis with a measure of the width of a bright star (using a parameter they developed, the Half Flux Diameter) on the y-axis. Since this is a linear relationship that forms a V shape, the precise point of optimal focus can be determined where the lines meet at the point of the V. The slope of the V's arms is a function of the focal ratio and pixel size, so once the software has characterized the optics, it can quickly find optimal focus after only a few images. FocusMax integrates with most commercial imaging applications and focusers. FocusMax can also accommodate input from a temperature sensor. This allows for automatic adjustment of focus based on temperature after the optics' response to temperature has been characterized. Once you have had the luxury of automated focusing, it can be hard to go back to any manual method.

For those without a computer nearby, many mechanical aids for focusing have been invented over the years, with focusing masks the most popular of these. Some of the simplest masks, like the Hartmann mask, consist of an opaque mask with three (usually triangular) holes. The mask is placed over the objective when the telescope is pointed at a star. As the point of focus is approached, the three shapes converge into one. This is an effective way to achieve focus, but it can be hard to judge the precise point of critical focus. Two newer and more complex designs remedy this issue, making it clear when you are approaching focus, and more importantly, when you have gone past it.

The Carey and Bahtinov masks were both created by amateur astronomers (George Carey and Pavel Bahtinov, respectively), and both generate an X-shaped diffraction pattern. With the Carey mask, two X's appear to be offset by a small angle. When the separation of the arms is equal, the optics are at optimum focus. The Bahtinov mask produces an X shape intersected by a 'crossbar.' Focus is achieved when the crossbar is aligned over the center of the X. Either of these masks, combined with an updated view of the camera's output, has improved upon previous methods enough to win the hearts of many imagers. You can create a custom template for your telescope online (try www.astrojargon.net for a Bahtinov mask generator), then cut it out of any thin, opaque material. Commercial, laser-cut masks are also available. Given the simplicity of making one, or the minimal cost of purchasing one, there is little reason to recommend older focusing aids.

Figure 51. A Bahtinov mask

Acquiring Images

The series of images in Figure 52 shows a magnified view of how focusing is achieved through an 8" f/8 Ritchey-Chrétien with a Bahtinov mask. When the long diffraction spikes are centered, the telescope is perfectly focused. The shorter diffraction spikes close to the star are from the telescope's vanes.

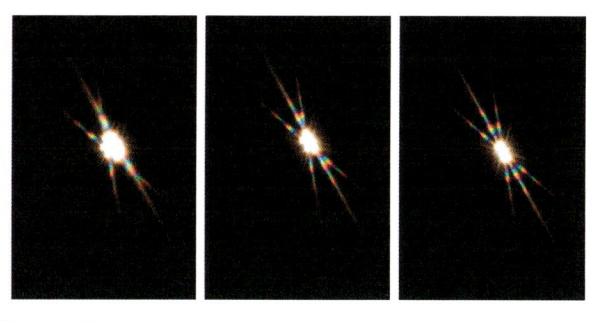

Figure 52. Achieving focus with a Bahtinov mask

Using focus masks is simple, but there are a few things to bear in mind. First, you must use a star for a target. Planets may be tempting because of their brightness, but since they are not point sources of light, they do not create the same clear diffraction pattern as a star. Focusing through the viewfinder of a DSLR is also not recommended, not only because the view is dim and small, but also because it is dependent on the accuracy of the reflex mirror's placement relative to the actual plane of the sensor. If your camera has live-view capability or you can connect it to a computer, a magnified view of the sensor's output is the best guide to focusing. If not, quick test exposures of bright stars can be made at different points of focus.

If you are using a go-to mount, it can be convenient to set focus during alignment, when you'll be pointing the telescope at bright stars anyway. However, if the telescope is not at thermal equilibrium with the environment yet, wait until it is before focusing.

It is crucial to use a quality focuser and to lock it mechanically once focus is achieved. Since 1/10th of a millimeter can make a difference in image quality, good focusers need to have exceptional machining tolerances. Any wobble, slop, slippage, or non-orthogonality (where the sensor plane is not perpendicular with the light path) will impact your image. Crayford-style focusers are great, but quality is not uniform. Always verify that locking the focuser did not alter the focus—this can occur, especially with cheaper models. High quality rack-and-pinion focusers can also provide excellent results.

Also, do not neglect to refocus. Some telescopes are very sensitive to temperature, so they require frequent refocusing as the air cools. Most filter sets that claim to be parfocal are not exactly, so unless you have verified their relative focus points, the best approach is to refocus whenever changing filters. Focus also shifts with altitude, so choose a focusing star that is at a similar altitude as your target object. As the object moves during the night, re-focusing may be required, and the effect becomes greater as an object approaches the horizon.

> For a *really* simple mask, try the Lord mask by Chris Lord. He points out that the same diffraction pattern generated by the complex Bahtinov mask is also generated by a simple Y-shaped opaque cut-out.

Autoguiding fundamentals

Ideally, an equatorial mount would track the rotation of the earth perfectly. Unfortunately, this is rarely the case, and there is no better way to reveal even small defects in tracking or polar alignment than a long exposure taken at a long focal length. With effective autoguiding, most tracking imperfections can be corrected, and it is one of the biggest steps toward getting images with pinpoint stars. Autoguiding primarily corrects for low frequency error, periodic or random, in the mechanics of the mount. It can also provide limited correction for imperfect polar alignment and the shift in apparent position due to atmospheric refraction (which is sometimes also accounted for by the mount's internal programming).

The basic idea of guiding is simple: watch a star somewhere in or near the field of view of the imaging sensor and send corrections to the mount so that star is held in exactly the same position. If polar alignment is accurate, usually only corrections in right ascension are necessary, either slightly speeding up or slowing down the mount's normal tracking. When declination corrections are necessary, most software will try to make all declination adjustments in the same direction, since there is usually some play or looseness in the gears.

Guiding used to be done by hand, looking through an off-axis eyepiece and sending corrections to the mount through the hand controller. Now, as with so many other things, a computer can automate this task better than any human. Autoguiding software can calculate the center of a guide star to within a small fraction of a pixel on the guide camera. This means that even long focal length telescopes can be guiding by small, cheap, and light telescopes of short focal length.

Getting your mount set up for autoguiding is surprisingly easy once you have the equipment, though there may be an evening or two needed to get the components "talking" to each other. Autoguiding results in a huge improvement in image quality for a small investment of time and money. One or two nights spent mastering it is one of the best investments you can make in image quality.

The light from the guide star can be captured several ways. The first option is known as off-axis guiding. Here, a prism or mirror is used to redirect some of the light directed just off the edge of the imaging sensor to a separate guiding sensor. Alternately, a smaller guide sensor can be placed right next to the imaging sensor. (This is only available on SBIG cameras, as they have a patent on this design.) The advantage of off-axis guiding is that you are guiding from the same optical path that you are imaging from. The off-axis methods are particularly desirable for telescope designs where mirror movement could be a problem, like SCTs. The primary issue with off-axis guiding is the difficulty of finding a guide star in the narrow field of view provided by a longer focal length imaging scope. And if narrowband filters are used, the light needs to be picked off in front of the filter to avoid losing most of the light.

The most commonly used method for autoguiding is to use a separate telescope and camera to guide. No light needs to be redirected from the imaging sensor, and a smaller widefield telescope can be used for a guide scope. The only potential drawback is that having separate imaging trains introduces the potential for microscopic shifts in alignment between the two telescopes, called flexure, which can prevent accurate guiding.

Thanks to great software and equipment, autoguiding is simple to implement. A specially designed camera is not needed; a common first-generation CCD camera will do a fine job providing a selection of guide stars in even the dimmest regions of the sky. (Note however, that webcams and planetary imaging cameras are not generally sensitive enough to make a good guide camera.) Even the guide scope can be of modest quality.

Each software package is different, but the fundamentals are the same. Once you have established a connection with the mount and the guiding camera, which usually require specific drivers, the software needs to calibrate. This will establish how much a given amount of movement in the mount will translate to on the sensor and in what direction. Calibration is generally necessary for each object you image, since changes in declination change the amount of mount adjustment necessary in right ascension.

Flexure

Differential flexure (usually, just referred to as flexure) is a change in the alignment between the guiding optics and the imaging optics. Even if the computer guides perfectly on the star it sees through the guide scope, if the imaging scope shifts relative to the guiding optics, the two cameras are no longer aligned. Elongated stars in the final image are a common sign of flexure. Successful autoguiding requires angular accuracy to within a tiny fraction of a degree, and even microscopic physical shifts between the imaging and guiding optical systems are enough to cause issues. Temperature change is a common culprit; as the night cools and metal contracts, the result can be flexure. Gravity can also cause flexure as the mount moves, especially if the guide scope is attached via non-rigid materials like plastic or wood.

The common points of vulnerability for flexure are the rings, dovetail, and screws that hold the guide scope. The focuser and the connection between the guiding optics and guide camera can also sag if not tightly secured. Avoid flexible materials like plastic wherever possible, and especially avoid wood, which warps with humidity. Tight metal on metal connections are preferred.

> First-order adaptive optics offer an improvement to standard autoguiding, allowing many corrections per second by using a small motor to shift the reflected view on a small mirror, rather than trying to move the whole mount. SBIG offers several AO-series adaptive optics units for amateur use.

Connecting computer to mount

There are several ways for a computer to communicate with the mount when autoguiding. Most mounts have a port that looks like a telephone jack, which allows direct control of the mount's motors. This is usually referred to as an autoguiding port, or sometimes an "ST-4 compatible" port, which refers to model number of the first standalone autoguider, developed by SBIG in the 1990s. The port is actually a standard RJ-12 (6 pin, 6 connector) telephone jack, and standard phone cables can be used (make sure it is labeled "6P6C"), but the specific standard for pin assignments and their correlation to direction is what the ST-4 established. Even this standard is not universal, however, with some mounts retaining the jack, but using slightly different pin assignments. Fortunately, most guiding software can figure out what signal to send for each direction during the calibration process. To connect to the autoguider port requires an interface box with an RJ-12 cable on one side and a USB cable to communicate with the computer on the other end. One standard interface box is the GPUSB

Acquiring Images

by Shoestring Astronomy. Most cameras specifically designed for autoguiding can communicate directly with the autoguiding port, eliminating the need for an additional USB cable and interface box.

Alternately, you can connect to many mounts indirectly through their hand box. This does not control the motors directly, but instead sends a signal to the microprocessor requesting a change in motor speed, which accomplishes the same thing with a few milliseconds of delay. This requires a special cable, typically a smaller jack (like the RJ11/14 connector), and each manufacturer has specific standards for their cable pinouts. The special cable connects with a serial controller on the computer (using the 9-pin trapezoidal connector known as a DB-9). Since few computers made in the past decade still include a serial controller, USB-to-serial adapter is usually needed. Mounts still use the old serial connection standard, despite the fact that few computers still include it, because it is far simpler and more reliable than USB.

The computer and mount also need a common language. Fortunately, there is a standard available here. ASCOM (AStronomy Common Object Model) is a freeware initiative that allows software to interface with a range of astronomical equipment, including mounts, focusers, observatories, cameras, and more. Typically, you can install the drivers for your specific mount or interface box when you install your autoguiding software.

Standalone autoguider packages are also available: the SBIG SG-4, the Celestron NexGuide, the LVI SmartGuider, the Orion Starshoot Solitaire, and others. These do not require a computer; they contain a camera and computer that can automatically lock onto a star and communicate with the mount. If you are already using a computer to capture images from the camera, these may be of limited value, but they can provide a very portable setup for those with DSLRs, and when they work well, their convenience is a huge benefit. Meade's LX800 mount even includes autoguiding capabilities built into the mount.

If you do choose to use a computer, the location of the computer will depend in your set-up. An observatory is the ideal, with everything sitting together in close proximity. For many amateurs, a semi-permanent backyard arrangement is as good as it gets, and it is usually desirable to keep the computer inside where it is comfortable. Figure 53 shows a typical setup like this, with one power cable and one USB cable running out to the outdoor equipment (hopefully in a weather-proof container). Note that the USB hub is a powered hub, which can be required to drive many guiding cameras, especially given the weak current provided by most laptop USB ports. Wi-fi also opens up a range of possibilities—consider leaving a computer outdoors, then using remote desktop software to control it from inside. Given the modest demands of most guiding and image capture software, a cheap netbook or even a thin-client computer can be left at the scope.

Software and settings

As of 2012 the most commonly used autoguiding software among amateurs is called PHD ("Push Here Dummy"), an excellent piece of free software from Craig Stark, a dedicated imager who also happens to be a professor of neuroscience. There are other options available, some included as part of larger image processing packages like Maxim DL, but for straightforward guiding, PHD is hard to beat, with

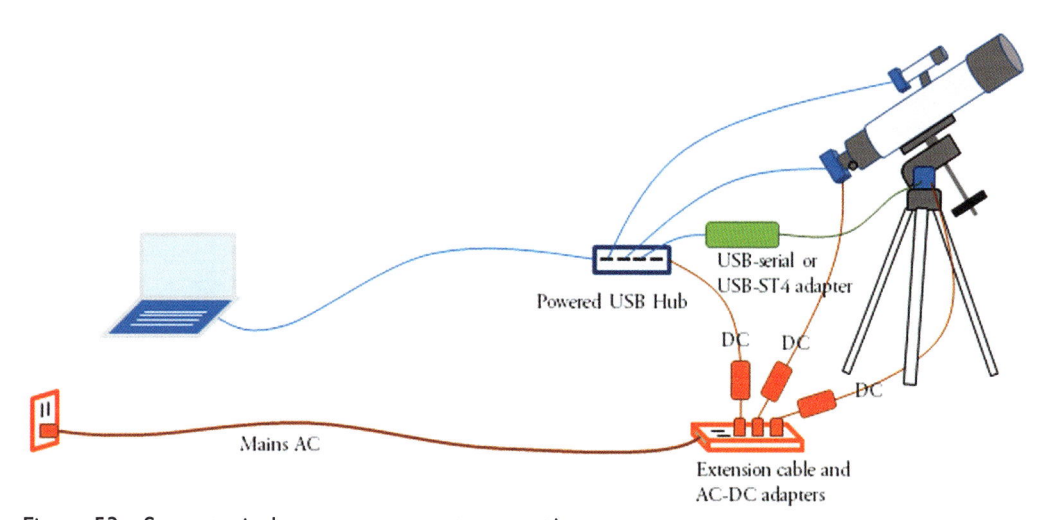

Figure 53. Some typical computer-mount connections

its simplicity and intuitive interface. (PHD, along with Craig's other software, is available at www.stark-labs.com.)

At a fundamental level, all autoguiding programs work similarly. First, the software must verify that it can 'speak' with the mount and the camera. Once it starts capturing images, either you or the software must find a star in the field of view to use as the point of reference: the guide star. The software can then calibrate itself to the mount by asking it to move in each direction for a set amount of time. By watching the shift in the guide star's position, it calculates how long a given guide command (which typically drives the mount at twice the tracking rate) will move the mount and in which direction. Once the software has connected and calibrated, it can monitor the guide star and send corrections to the mount that will return it to the original spot if it shifts. While corrections are sent after the star has moved, by repeating the monitoring and corrections in a loop, the star can be held within a very small range. The size of this range will determine how tight stars can be in the exposures.

Figure 72 shows PHD's main interface during an active guiding session. The guide star is centered under the green crosshairs. Recent deviations in RA and declination are charted in the additional open window. The icons along the bottom from left to right are used to: connect to the camera, connect to the mount, display an updated view from the guide camera, start calibration/guiding, and stop guiding.

A star's image is spread across many pixels, so the software calculates the star's center to sub-pixel precision. In fact a sort of 'center of gravity' approach is used, called centroid calculation, that weights each pixel by its brightness to improve accuracy even with stars that only span a few pixels. Because of this, perfect focus is not necessary to achieve stable guiding. In fact, very slightly defocused stars are recommended for optimal guiding.

When moving to a new target, the guiding software will need to be recalibrated. Differences in declination cause the sensitivity to right ascension adjustments to change. To shift the position of an object near the celestial pole requires a lot of RA movement by the mount, whereas an object near the meridian will appear to move a lot with even small RA adjustments. For objects close to the pole, successful calibration may require you to increase a step size setting to induce enough movement. Meridian flips can also require recalibration, since the effect of a change in declination is reversed, though some software can handle this with a setting change rather than a new calibration.

Sloppiness in the mount's gears must also be overcome. Whenever the mount changes direction, it takes a fraction of a turn to re-engage the gears. This is called backlash, and in order to avoid it, most software will try to only send guide signals in one direction, especially for the right ascension motor. This keeps the gears engaged, allowing for predictable motion. The settings of your software will allow you to control how to account for backlash.

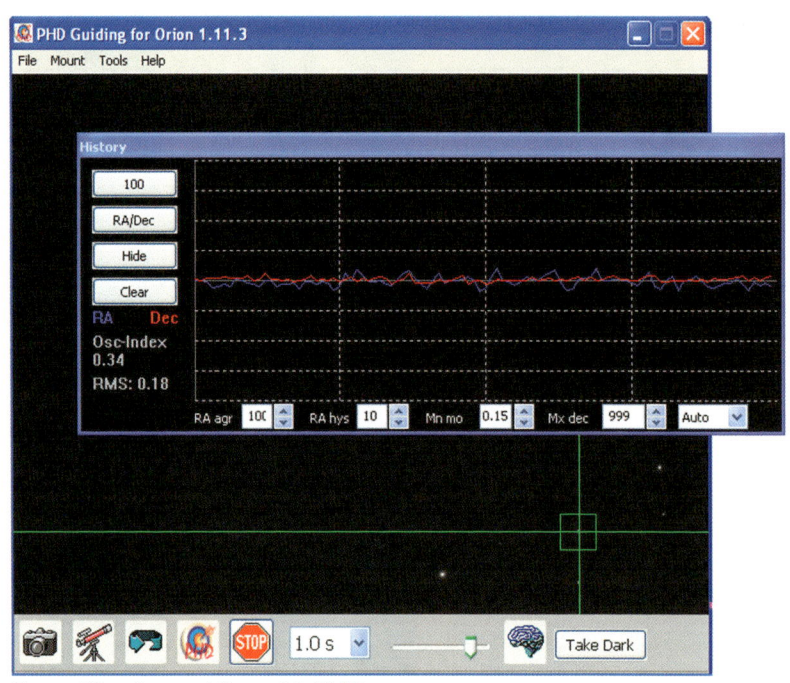

Figure 72. Autoguiding with PHD

Software settings can also prevent the mount from "chasing the seeing." Atmospheric turbulence causes a star's image to shift constantly. If the exposures are too short and the guiding too aggressive, needless adjustments will be made to try to correct for these atmospheric effects. Guide camera exposures on the order of 0.5 to 2 seconds are insensitive to twinkling of the guide star.

Choosing a scope and camera for autoguiding

Because of the accuracy of centroid-based autoguiding software and the sensitivity of CCDs, long focal length telescopes can be guided effectively by guide scopes that are much shorter in focal length and smaller in aperture. Perfect optics are not needed, either. Even the lowly 50 mm finder scope is perfectly usable for this purpose, not to mention very affordable. (Every image in this book was guided with a 50 mm finder.) Another benefit of using a very small telescope like a finder scope is that it adds very little mass to your setup, which can also help reduce flexure. Before purchasing a separate telescope specifically for guiding, consider if you might be better off with a much more affordable option. There are even marketed products like the Kwiq Guider that use standard 9×50 mm finders for guiding.

The ideal guiding camera is sensitive and small. Monochrome cameras have better sensitivity, since they have no filter matrix on the sensor, though color cameras can be used with longer exposures or brighter stars. Smaller chips are fine, especially with short focal length guide scopes. It is not necessary to spend a lot for a guiding camera.

> Modifying an old finder for guiding is a very simple do-it-yourself project, using materials found around the house or from your local home improvement store. The finders included in telescope packages are not usually designed to accept eyepieces, though some higher-end models are. The sensor may have to be positioned closer to the objective lens than the tube will allow, so it may be necessary to cut the tube and add your own adapter to mount the camera. Small pipe fittings or 1.25" eyepiece adapters can work for this. The best finder scopes have helical focusers, while basic finders focus by unscrewing the objective lens and locking it with a screwable ring. Either way, it is best to fix focus in place once it is achieved.

9   Setup and accessories

Reducing setup time

Sometimes the greatest impediment to imaging is simple inertia. You're on the couch, the skies are clear, but it's cold, and setting up seems like such a chore. This is the sort of situation that drives many to build a backyard observatory, but if that's not a viable option for you, you can still cut your setup time dramatically by streamlining your setup. The easier it is to set up, the more images you'll take, the more practice you'll get, and the better your results will be.

Polar aligning the mount is usually the most time-consuming step in getting set up for the evening, so it's wise to save time here if you can. Recent mounts from Meade, Celestron, and iOptron have computer-assisted polar alignment procedures that can be much faster than the traditional drift method. A simple step toward even faster alignment is to have a fixed place to put your tripod or portable pier. If your preferred spot is in your yard, place a three small bricks into the ground where your tripod legs stand when polar aligned. If you observe from a concrete or asphalt area, you can chip small divots or simply mark the ideal spot for each leg. Once you have your spots marked, you can leave the tripod legs adjusted so the mount is level when they are placed on their spots. This will not save you from having to properly polar align the mount each night, but it will at least get the alignment process off to a faster start.

The next step toward a permanent arrangement is to move from a tripod to a pier. Typically, these are constructed from concrete, but creative astronomers have also used wood or metal. A properly constructed pier holds its position precisely over time and doesn't vibrate or shake easily. This stability allows for the potential to polar align the mount only periodically, instead of with every imaging session. Each pier is a custom construction, and they can require a substantial effort to build. A concrete footer is typically poured under the pier that is several times larger than the pier itself, and everything must not only be extremely stable, but the mount attachments must be accurately leveled and weather-proof. It's a serious undertaking, but if you have the skills and the space, you can be set-up and imaging in minutes every clear night.

Once you have a permanent pier, you're only a few steps away from a full-blown observatory. This can be a roll-off construction, where the roof slides off, or something more complex. There are pre-molded plastic observatories available as well. Entire books are dedicated to the creation of an observatory, as the sky (or your homeowners' association) is the limit here.

There are many other cloudy night projects you can do to speed up the setup process. All the necessary power cords, USB cords, and other connectors can create a mess. Taping them down exactly where they are needed can help. A USB hub can be strapped directly to the mount (the eyepiece holder makes for a good spot.) Even better, create a ventilated box to contain all the needed AC adapters, USB adapters, hubs, and cables. Everything can remain connected inside the box, minimizing the chances of loose connections, and greatly simplifying setup. Work the kinks out of your software and hardware on non-imaging nights, or during the day. Wrap cables together into bundles and tape over connections that need to stay together. All of these things will reduce frustration when it's dark and your hands are numb from the cold.

Figure 54. Bricks that line up the tripod are a simple time-saver

Other important equipment

Like most hobbies, astroimaging offers endless accessories and upgrades. Upgrade Madness (a sister disorder to Aperture Fever) can strike imagers at any time: when they see a sensor with a better quantum efficiency, a mount with better tracking, or software with new features. All of these mounts, scopes, and cameras can be combined in different configurations, frequently requiring some sort of adapter

Acquiring Images

or retrofit. Connecting everything together requires care, precision, and lots of adapters.

Dovetails, rings, and other mount accessories

There is a wide range of accessories available to attach a telescope to an equatorial mount. The telescope is typically held by metal rings that are screwed onto a dovetail plate (or "bar"), which is in turn held by the mount. Clamshell rings exactly match the outer diameter of the telescope, opening with hinges to hold it securely all around. Universal rings are available that hold smaller tubes between three plastic-tipped screws, but clamshell rings are preferred as they hold more securely.

Rather than screwing the telescope or its mounting rings directly to the mount, most are instead attached to a long metal plate with a trapezoidal profile called a dovetail plate. This plate is gripped by the mount in a clamp called the saddle. This standardized dovetail plate-saddle arrangement allows any brand of telescope or rings to attach to any mount. It also makes it easier to balance the declination axis of the mount by sliding the dovetail forward in the saddle. The industry has settled into two major standards for dovetails, one from Vixen, the other from Losmandy. The Vixen-style dovetail is the smaller of the two, with a maximum width of about 1.7 inches (43 mm) that converges to a minimum width at the top of around 1.4 inches (35 mm) at about a 12 degree angle. Losmandy plates (known as the "D series" plates) are wider and flatter, with a 3 inch (76 mm) wide dovetail groove cut out of a 4 inch (102 mm) wide plate. They are typically about 0.5 inches (12.7 mm) thick, but this can vary.

Typically, smaller mounts use the Vixen standard while larger mounts use the Losmandy standard. Dovetail bars are available in any length to fit a range of telescope sizes. The spacing of holes in the bar varies between manufacturers, as do the size of screw used for the rings.

Adapters are available to convert between the two dovetail standards. These are helpful accessories when using a range of scopes with one mount. Figure 55 and Figure 56 show examples of the two common types of dovetail plates.

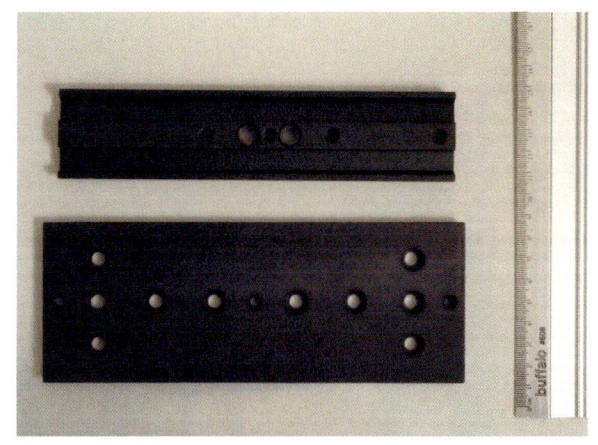

Figure 55. Vixen-style (top) and Losmandy-style (bottom) dovetail plates

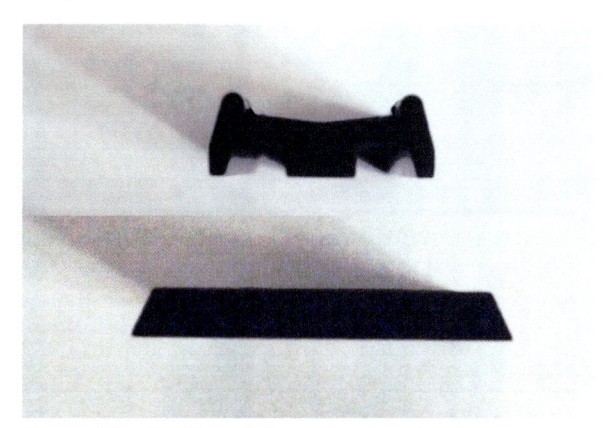

Figure 56. Profiles of the same two plates.

Power supplies

Ideally, an outlet provides power when imaging from home, but sometimes the darkest skies are far from utility power. Imaging gear consumes a lot of power. Mounts draw very little power while tracking, but slewing requires significant current. Cooled CCD cameras need surprising amounts of energy, and don't forget the laptop, USB hubs, and anything else in the imaging chain. All of these require direct current (DC) at different voltages. Separate battery packs can be used to power each item, or a power supply can generate the necessary voltages for multiple items with a single battery.

When using battery power, remember that most rechargeable battery packs fail quickly in cold weather. Even if you're imaging from home with a DSLR, powering the camera via an adapter is preferred. Not only does this ensure you don't run out of power in the middle of an imaging run, but it helps keep thermal noise down by removing the battery's warmth.

The do-it-yourselfer can find a plethora of projects in this area. For instance, a laptop power supply (cheaply available online) can power a mount if the proper peak current specifications are met and the proper plug is retrofitted to match the mount's power jack. When choosing a power supply, note that peak current requirements for mounts can be surprisingly high. Several amps are needed as a mount's motors engage. More sophisticated projects involve creating a dedicated rechargeable power supply driven by a deep-cycle battery. If you are looking for an off-the-shelf solution, there are commercial power supplies designed for the appropriate conditions and voltages needed by amateur astronomers.

Dew prevention

The goal of dew prevention is to keep condensation from forming on a lens or mirror surface. This can be accomplished in most situations by raising the temperature of the optics only a few degrees above ambient.

A dew cap or shield is automatically a part of most refractor designs, but they can easily be added to SCTs and other designs. By holding a pocket of air close to the telescope, dew shields slow down the equilibration of the optics, keeping them slightly warmer as the ambient air cools. This can prevent or at least delay condensation when the temperature drops below the dew point.

Electric dew heaters provide more active measures. These are straps containing resistors that attach around the telescope's tube or dew shield. They connect to a controller that pulses current through them, generating a small amount of heat. Each controller can typically drive several straps (one for the imaging scope, one for the guide scope, etc), and the power level is usually variable. Producing too much heat can create tube currents that interfere with seeing, so set them to the lowest effective setting.

As an alternative, you can use a small dew-gun or hair dryer to occasionally warm up the problem optics. The downside here is that they can draw a lot of current if you are on battery power. More importantly you have to stay awake and near your setup, which doesn't allow for automated imaging (which allows for sleep, which allows for a day job, which pays for more equipment...). In a pinch, you can even strap a small chemical hand warmer pouch near the dew shield, though you have to be careful with this approach not to apply too much heat.

USB hubs and extension cables

Connecting cameras, focusers, serial adapters and other USB items for imaging is more complex than most USB peripherals. Cameras have high power demands, and the cables are frequently run great distances. Powered hubs and "active" cables are necessary in these circumstances. USB 2.0 is the minimum speed required for most applications. Unfortunately, USB cables can only reach a distance of five meters before they need active signal boosting or repeating. To reach more than 5 m, up to three active USB cables can be linked together. Going beyond three links generally requires a powered USB hub, since voltage attenuation becomes a problem after 15 meters. There are also USB-to-ethernet adapters than can facilitate longer distances, but test them with your specific hardware as there are many complaints of compatibility issues.

Few mounts use USB directly, since serial communication cables are simpler and less error-prone. To interface with a mount for autoguiding, either a serial connection or ST-4 style guideport can be used. Either one will require an adapter to connect with a computer through the USB bus. To connect to a mount via an ST-4 style guideport, either your camera must be designed specifically for guiding and thus provides a direct output to the mount (as with Orion's Starshoot Autoguider and similar products), or you'll need a product like the GPUSB from Shoestring Astronomy (www.shoestringastronomy.com) designed to facilitate communication to the mount via your computer's USB port.

To connect via a serial connection, USB-to-serial adapters are readily available, but the quality varies widely. To connect with most mounts via a serial connection, not only is a USB-to-serial adapter needed, but a special cable is also needed to convert the DB-9 serial connection to a smaller connection on the mount (usually RJ-11 and usually on the hand box). Note that while they are physically similar, the pinouts on Meade and Celestron serial adapter cables are not the same.

10 Filters and Narrowband Imaging

Filters for imaging

There are three major color sets used for imaging with a monochrome camera:

- LRGB (Luminance, Red, Green, Blue) filters are used for capturing standard color images. Luminance filters block only UV and IR, while the RGB filters pass about a third of the visual spectrum each.
- UBVRI (Ultraviolet, Blue, Visual, Red, Infrared) filters are used for photometry, which is a subject for another book. They are not used in aesthetic imaging.
- Narrowband filters are used for imaging specific astronomical emission lines. They pass bandwidths less than 10 nm wide.

This chapter focuses on narrowband imaging. The processing methods for combining monochromatic images into a color image are the same for narrowband and LRGB. These are covered in the next part of this book.

The spectrum of our sun and most stars is nearly continuous across the visible spectrum. A few astronomical objects, however, contain ionized gas that emits light at only specific wavelengths. These gases emit light when an electron falls from a higher energy orbital to a lower energy one. Because the energy levels are discrete (or 'quantized'), each electron state change for each gas produces a specific wavelength of light. By using filters that pass only the light within a very narrow band of wavelengths around one of these emission lines, we can image these emissions while excluding the rest of the spectrum. This is known as narrowband imaging. For specific objects, this leads to a quantum leap (excuse the pun) in signal-to-noise ratio. In fact narrowband imagers have the luxury of capturing quality data from the most light polluted locations, and even during a full moon, greatly increasing the number of available nights for imaging. In a way narrowband imaging offers dark skies for everyone, at least when imaging emission objects.

The interference filters used in imaging are not like the dyed-glass filters used in daylight photography. Dyed-glass filters pass a low proportion of light, and the shape of their transmission curve shows a gradual transition into and out of their main pass band. Interference filters on the other hand usually pass more than 90% of the light within their pass band, and they have steep boundaries between the wavelengths they pass and those they do not.

Because so little light passes through the filter, fast optics and long exposures are required. F/5 optics or faster are common, and even then, exposures usually exceed 20 minutes to get useful data. Most of the best narrowband objects are also large, providing plenty of options for those with telephoto lenses and short focal length telescopes.

The final images produced by narrowband imaging look markedly different from standard broadband images. Many nebulae are beautiful even in monochromatic narrowband images where fine structural detail is revealed. Emission nebulae stand out more against the background stars when viewed through narrowband filters, since only a small portion of the light from the broadband-emitting stars are captured. Further, by collecting light from specific emission lines and assigning them to different colors in a false-color image, we can visually separate the aspects of a nebula's structure that result from each gas.

Narrowband filters work best with monochrome cameras. If you are using a DSLR or one-shot color camera, any narrowband filter will be in front of the sensor's existing color filter array. For instance, an H-alpha filter will pass only that specific red part of the spectrum, thus the green and blue photosites will not receive any light at all, wasting two-thirds of the sensor. This is not to say that it is impossible to use narrowband filters with color-array sensors, but it is less than ideal, and given the costs of the filters, consider purchasing a monochrome sensor. Because of the long exposure times required, thermal signal can also become a significant issue without active cooling.

Figure 57 and the table show the most common emission lines and their wavelengths. The vast majority of narrowband images use only the three most common filters: H-alpha, OIII, and SII. Filters are available for other less common emission lines, but these lines are not as bright and are rarely used by amateurs.

H-beta emissions generally come from the same sources as H-alpha, but since H-alpha is about three times brighter, it is the more commonly used filter. If you only own one narrowband filter (which is understandable given the cost), H-alpha is the best choice. Since so many of the most

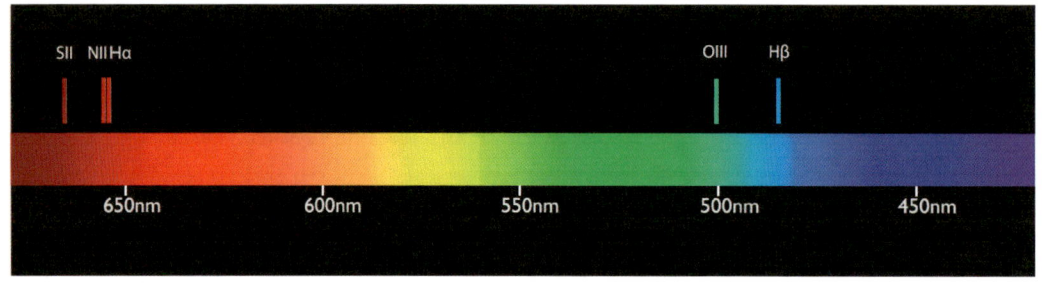

Figure 57. Commonly imaged emission lines in the visible spectrum

beautiful nebulae emit primarily at this line, it can be used alone or with RGB broadband filters. When used alone, it produces beautiful monochromatic images of red emission nebulae. Used with RGB, the addition of an H-alpha channel can provide contrast that emphasizes the fine structures in these objects.

Narrowband filters pass only a very small range of wavelengths, called the bandwidth. Where a broadband filter like a red, green, or blue imaging filter pass a range around 100 nm wide, the tightest narrowband filters pass only a 3 nm bandwidth. Nearly all narrowband filters have bandwidths smaller than 10 nm. Since emissions are at a very specific wavelength, the narrower the bandwidth, the more unwanted light is excluded, which improves the signal-to-noise ratio. This also dramatically reduces the effects of light pollution, as shown in Figure 58. Without a filter, the total energy of the skyglow (top, shaded red) across the spectrum dwarfs that of the nebula. With a narrowband filter centered on the nebula's emission peak (bottom), the total amount of light reaching the sensor is far less, but the proportion from the object is much greater.

The spectrum of moonlight is nearly the same as that of the sun it reflects, but the same Rayleigh scattering that makes the daytime sky blue also affects moonlight. A moonlit sky can be thought of as a darker version of the daytime blue sky. Since atmospherically scattered moonlight is brightest at the blue end of the spectrum, it has more effect on imaging at the OIII and H-beta wavelengths than the H-alpha and SII wavelengths. For filters of the same bandwidth, an OIII exposure in moonlight will show a brighter sky background, degrading SNR more than an H-alpha or SII exposure on the same night. Sequence your imaging so OIII falls on the darkest night or use an OIII filter with a narrower bandwidth.

Any difference in the thickness of filters will cause a shift in focus, so buy a parfocal set of filters. Though no filters will be perfectly parfocal, for faster focal ratios a parfocal set can help you avoid unnecessary refocusing. Verify the focus shift after each filter change at faster focal ratios. (This applies to narrowband or broadband color imaging.)

Emission line	Wavelength	Description
Sulfur-II	672.4nm	A deep red emission line that is typically dimmer than H-alpha.
Nitrogen-II	658.4nm	NII is so close to H-alpha that all H-alpha filters pass at least some of this signal. Bandpasses of ≤3 nm are needed to separate them, so NII filters are rarely used.
Hydrogen-alpha	656.3nm	A deep red emission from ionized Hydrogen. This line is dominant in emission nebulae and the HII regions of galaxies. H-alpha is almost always the brightest emission line.
Oxygen-III	500.7nm	A green-blue emission line from ionized Oxygen. It is typically dimmer than H-alpha.
Hydrogen-beta	486.1nm	A blue emission line from Hydrogen, representing an electron transition between higher orbitals than H-alpha. It is far dimmer than H-alpha. (Note that this is not the source of blue light in reflection nebulae.)

Acquiring Images

As we'll see, narrowband images also require more careful processing. The histogram can remain a thin spike, even after long exposures. To tease apart these levels, keep the data in the highest bit depth image files possible when stretching. Narrowband data also presents a dilemma when integrating the color channels in a false color image. Some artistic license is usually taken when assigning narrowband data to colors. The most commonly seen mapping is the 'Hubble palette' (also known as the wavelength-ordered palette), where SII, H-alpha, and OIII are assigned to the red, green, and blue channels respectively. Even after assigning narrowband channels to standard colors, further tweaking is usually necessary to balance the channels and create a pleasing result. All of this will be covered in the next part of the book.

The H-alpha signal is far brighter than the other emission lines for most objects, so you must typically gather several times more OIII and SII data to permit the stretching necessary to bring it to the same level as H-alpha. Nearly all H-alpha filters have a bandwidth wider than 3 nm, so they also capture the nearby NII line, further strengthening the impact of this filter's data in the image. While it's tempting to go for more H-alpha data because the exposures look so nice, more exposure time should be dedicated to the other two channels so they have comparable SNR to the H-alpha channel.

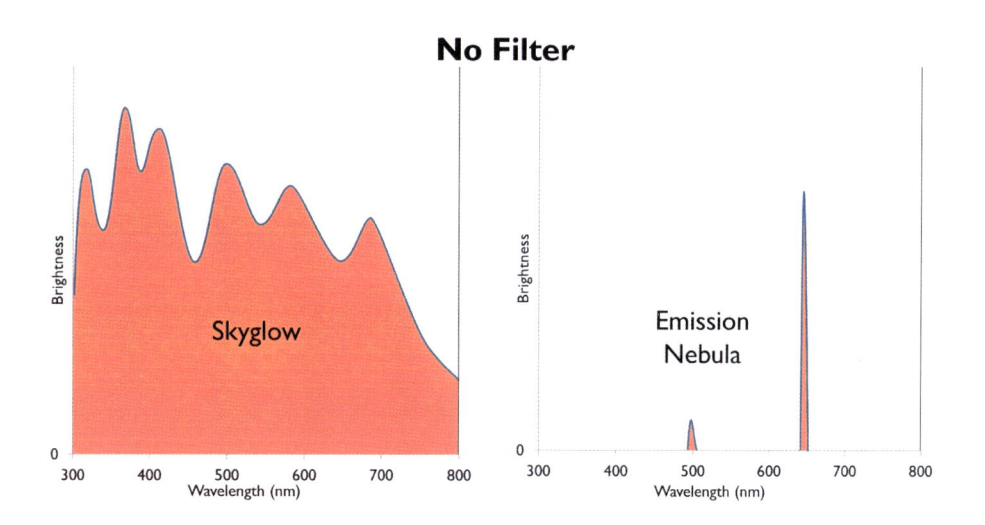

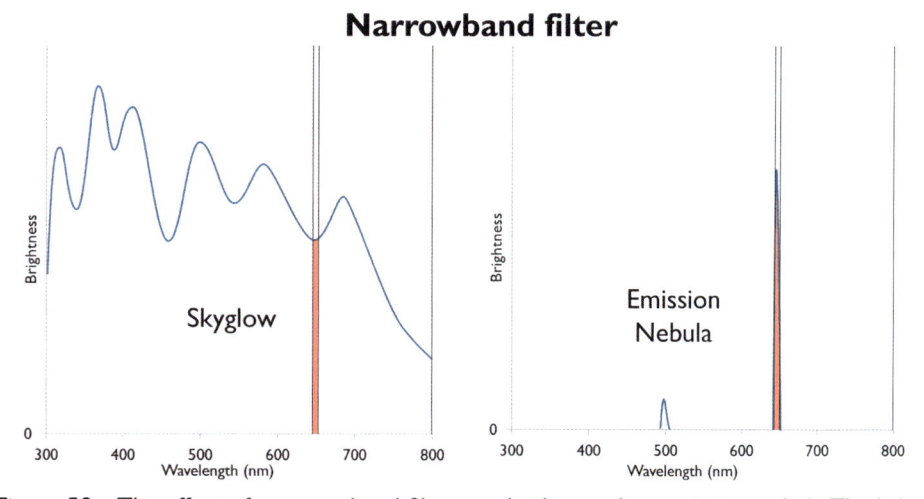

Figure 58. The effect of a narrowband filter on skyglow and an emission nebula. The light reaching the sensor is shown in red.

> The names of the emission lines can be confusing. Because of its importance in the discovery of the atom's structure, Hydrogen's emission lines have unique nomenclature from atomic physics. The most commonly seen terms in astronomy are the Balmer series of Hydrogen spectral lines. The first in the series, H-alpha, denotes the emission line generated when an electron falls from the n=3 quantum state to n=2. H-beta indicates a transition from n=4 to n=2, and so on.
>
> For other elements, the language of spectroscopy is used. Here, the elemental symbol is followed by a Roman numeral describing its ionization state: I is for a neutral atom (the consequence of Roman numerals not having a zero), II for the singly ionized state, III for the doubly ionized state, etc. For instance OIII describes doubly-ionized Oxygen (two missing electrons), and the SII emission line is singly-ionized Sulfur (one missing electron).
>
> Mixing nomenclature, one source of H-alpha emissions are what are known as the HII ("H-two") star-forming regions in some galaxies. The HII is the singly ionized state of Hydrogen when it is a free proton—Hydrogen cannot be doubly ionized, as it has only one electron in its neutral state. When this proton captures an electron to become neutral Hydrogen, the electron falls through quantum states, emitting photons as it does so. One of these is the H-alpha Balmer line.

Light pollution filters

For one-shot color cameras, broadband light pollution (LP) filters are available to block portions of the spectrum produced by common outdoor lighting, allowing the rest of the light through. The sodium and mercury vapor lamps commonly used for outdoor lighting do not produce a continuous spectrum of light, so filters that reject the brightest wavelengths they emit can be used to reduce the effects of light pollution when imaging. Some light pollution is full spectrum, as is moonlight, so the filters can't help much there, but for urban and suburban imagers, they can improve SNR.

While it may seem paradoxical that dimming the overall view is a good idea, the principle is the same as we saw with narrowband imaging. For those in light polluted areas, skyglow is the source of most of the signal, and thus most of the noise. LP filters do eliminate some of the light from astronomical targets, but the reduction is proportionally much smaller than the effect on skyglow. This increases the proportion of light captured that came from the target versus that from skyglow, and it also allows you to expose longer before saturation occurs on the sensor. So even though these filters reduce the overall brightness of the image substantially, by selectively reducing the skyglow more, they result in an improved signal to noise ratio (and thus a higher quality final image). The worse your local light pollution, the more improvement these filters can deliver. Targets that emit only a few discrete wavelengths, like emission nebulae, see substantial improvement, while star clusters, galaxies, reflection nebulae, and other broadband targets may not see as much change.

> Older orange low pressure sodium lamps produce nearly monochromatic light around 589 nm. Newer high pressure sodium lamps produce a whiter light composed of several colors, but most of the energy is still in four broad peaks (around 570, 590, 600, and 630 nm). Mercury vapor lamps also produce light that is concentrated near four specific wavelengths (405, 436, 546, 578 nm). Other outdoor lamps, like metal halide, produce a nearly continuous spectra that cannot be selectively filtered as well.

Acquiring Images

11 Taking the exposures

Controlling the camera

Now that you have obtained and set up all of your equipment, it's time to actually take the exposures. Whether you are using a dedicated astronomical camera or a DSLR, it is best to control your camera from a computer (actually, it's the only option for dedicated cameras). This allows you to automate your imaging runs and view the exposures as they are taken, which can provide a valuable warning if there are tracking problems or the composition needs to be adjusted.

If you're a DSLR user, astroimaging usually requires exposure times longer than the 30 seconds most cameras can time internally, so you'll need to use the "bulb" setting and control the exposures either through the remote shutter release port or through the USB port. Older DSLRs required two or more cables, since the USB could capture exposures, but not trigger the shutter (for Canon, these were the models before the 450D/Rebel XSi). For basic automation of DSLR exposures, handheld programmable timers are available for under $40. Typically, you can set a delay before starting the first exposure, the length of the exposures, the interval between exposures, and the number of exposures to take.

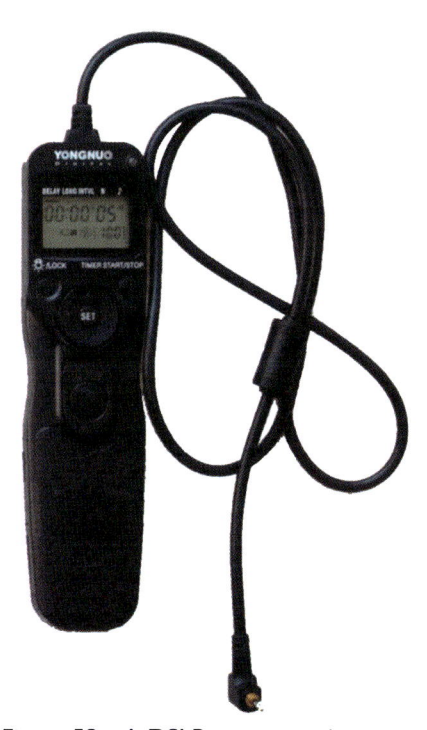

Figure 59. A DSLR exposure timer

There are many software options for controlling your telescope and camera from a computer. Most cameras come with basic software to control exposures. Canon's EOS Utility software is a good example of a simple program for DSLRs that is sufficient for most imagers, but new third-party programs like Backyard EOS and Astro Photography Tool (APT) go beyond these simple functions to allow different exposure times in a session, dithering of exposures, focus assistance and more. CCD cameras nearly always come with software to control exposures, but they vary in terms of complexity and integration with other software.

Another option is to use one of the dedicated astronomy packages to control your camera. They combine a range of features like camera and filter wheel control, image processing, planetarium software, and even observatory control. Among the cheapest and easiest to use is Nebulosity from Stark Labs (now at Version 3, $80 at time of print), which offers basic image processing in addition to camera control. Comprehensive packages like MaximDL, Images Plus, and CCDSoft add an array of sophisticated features, usually at much greater cost.

Plan your imaging sessions in advance. Get set up early, so you can maximize the hours of darkness. Check the weather forecast. Know when the object is best placed relative to the moon or your local light pollution sources so you can reduce the impact of gradients and extra noise. As an object approaches the horizon, atmospheric distortion and light scattering seriously degrade image quality, so don't expect to collect usable data in that area of the sky. If you are taking filtered exposures, either narrowband or LRGB, consider the order of filters you apply. Take exposures filtered through the shorter wavelength filters (blue, OIII, H-beta) when the target is highest in the sky, and capture the red end of the spectrum when it is lower. Also consider the proximity of the moon with these same filters, as moonlight scattered by the atmosphere is brighter toward the blue end of the spectrum.

> The term "bulb" is a holdover from the days when pneumatic shutters were operated by an air bulb that would hold the shutter open as long as the photographer squeezed it.

83

Choosing exposure duration and gain

If you had a noise-free camera with infinite well-depth, it wouldn't matter if you took a single 60-minute image or combined 3600 one-second images. We know that in reality of course, it does matter. Real cameras add a small amount of read noise to each exposure. To minimize that read noise, you'd ideally want to divide your total imaging time across as few exposures as possible. Unfortunately, many factors limit us to shorter exposures.

Any sensor's photosites saturate at a certain point, failing to register additional photons after the well is full (of electrons), clipping highlights in the image. Thus the brightest parts of the image—whether stars, a galaxy core, or nebulosity—set a maximum on exposure length. In general it is best to shoot the longest possible exposures before the brightest pixels in the image are at risk of being clipped. The histogram for ideal subexposures would show the long tail on the right of the main histogram "hump" trickling off toward the right edge, but stopping just short of it. Exposures like this fit the full dynamic range of the scene into the sensor's well capacity, avoiding clipping the brightest areas while still capturing the faintest areas, which will need to withstand the most stretching later in post-processing. If the dynamic range is too great to for sufficient data to be captured in the dimmest areas without clipping the brightest areas, two sets of exposures may be necessary to adequately capture both. This is why larger well capacities are a benefit: they provide greater dynamic range. It is also another reason skyglow is such a problem. Since skyglow is subtracted from the final image, but the photons must still be collected, it reduces the available dynamic range for target data.

In many cases however, the sensor's full well capacity is not the rate limiting step for exposure time. Practical limitations come into play before that time is reached. Wind gusts, mount or guiding errors, or passing clouds can all ruin an exposure. Dividing the total exposure time into shorter subexposures helps mitigate that risk. The calibration processes also works best when there are many exposures to work with. For instance, using any of the "sigma clipping" methods to average the exposures will discard any outlier values, those that are more than a couple of standard deviations from the mean. This helps exclude the effects of passing airplanes, cosmic ray strikes, or other disturbances, but it only works when there are enough exposures to calculate a meaningful standard deviation.

We also can't swing too far in the other direction and take exposures that are too short. Small sources of noise like read noise and quantization error have a noticeable impact when we have too little data in each exposure. So where is that perfect balancing point between too long and too short?

As a rule of thumb, the image histogram is a helpful guide for finding the minimum useful exposure time. The left edge of the main hump, which represents the background, should be fully separated from the left edge of the histogram. This way, the dimmest signal in your image—the left edge of that hump—is likely to be significantly higher than the level of read noise. (The left edge of the histogram near zero is a good approximation for the *magnitude* of read noise, but it is part of the signal at every level of the histogram.)

Though read noise is very small relative to the total dynamic range of the sensor, it has a great impact on the faintest areas of the image. For these pixels, the read noise can be a significant fraction of the signal—in other words, the SNR is low. This leads to a grainy appearance that no amount of smoothing in post-processing can accurately repair. The further to the right you can move the hump, which contains most of your signal, the less impact read noise will have on the dimmest areas. But it's a balancing act; you still have to take care not to clip any data in that thin tail that extends rightward from the hump. There is only so much dynamic range that can be captured in one exposure, and the exposure duration determines how it is allocated.

There is abundant amateur literature on the internet concerning the calculation of a precisely optimal, or at least minimum, subexposure duration. Generally, the stated goal of determining such a value is to ensure that the exposures are long enough that read noise is not a significant source of noise compared to the shot noise from the target object. This usually involves picking an arbitrary proportion (usually 5%) of the total noise that can come from read noise. Given the read noise of a camera, the brightness of the dimmest area to be captured, and the brightness of the skyglow, one can calculate a theoretically ideal exposure time. Any such calculation is specific to the selected proportion of read noise, your sky conditions, and possibly your choice of object.

While it can be a good exercise to stimulate thought about the sources of noise in your images, don't get too caught up in calculating a theoretically optimal exposure time unless you have a rainy weekend to spare. There is a pretty wide amount of leeway around optimal exposure time before you'll see a significant effect on the final image quality, especially if every other aspect of your image capture and processing are not already optimized.

Take exposures that are as long as are practical for your setup, and use experience as your guide. If the histogram shows some clipped highlights, back off the time a little. For most skies and most focal ratios, typical broadband subexposures range from one to 20 minutes. Those shooting at fast focal ratios, with limited well capacities, or through heavy light pollution would be on the shorter side of this range. Because of the greatly reduced number of photons, narrowband imaging requires longer subexposures, frequently in excess of 20 minutes. Mount quality and alignment accuracy start to become more of a limiting factor here.

As for the gain setting, most dedicated astronomical cameras have fixed gain, simply mapping the full well capacity to the number of bits available from the ADC. DSLRs present gain as a sort of artificial construct in the form of an ISO setting, designed to make their use similar to film cameras. Though most current DSLRs have full well capacities of over 40,000 electrons, their ADCs and gain circuitry are not quite as good (read: noisier) as those in astronomical cameras. The optimum gain setting for a given camera depends on many factors. Some are inherent to the camera, like bias signal, ADC bit depth, and full well capacity. Others are about how you use it: bright skyglow, fast optics, or long exposures can raise the signal floor, eating into the usable dynamic range, making high gain settings less useful. It may take some experimentation to find it, but a good place to start is ISO 800 or 1600 for DSLRs. For current cameras with 12-bit ADCs, these settings put you near 'unity gain' (where one electron corresponds to one ADU in the raw data) without reducing the dynamic range available for highlights. Check the internet to see what other imagers with your camera are using and why. The ISO settings that are clearly to be avoided are the "in-between" ISO settings, like 250, 500, etc. These are usually the result of use two-stage gain circuits that add noise twice.

As noted before, there is nothing special about the precise number of exposures taken. The SNR (considering only shot noise) does indeed fall by integer fractions as you square the number of equal-duration exposures, but that does not mean that you need to take exactly 4, 9, 16, 25, etc. exposures. The calibration calculations are done to many decimal places, so any number of exposures will do—Fibonacci numbers, Mersenne primes, lucky numbers, whatever you are able to capture! Nor does the number of calibration frames required relate to the number of lights. You want as many of both as you can get. The same "more is better" rule applies to both lights and calibration frames: the greater the total exposure time, the lower the noise, though with diminishing (logarithmic) returns.

> To match the human eye's wavelength response, DSLRs have an infrared filter in front of the sensor. Unfortunately, this filter blocks most of the light in the H-alpha emission line at 656 nm, making it difficult to capture images of emission nebulae or the HII star-forming regions in many galaxies. This filter can be replaced with one that doesn't cut out the H-alpha line, either via a commercial service or as a do-it-yourself project.
>
> If you have a DSLR, filter replacement is worth considering. It will help reveal emission nebulae that would otherwise be nearly invisible to your camera, and it improves the view of many planetary nebulae and galaxies as well. That said, it is not cheap, it voids the warranty on your camera, and subsequent daylight photography requires an additional filter and/or custom white balance adjustments. It is a tradeoff that many people have made, but it is certainly not a necessity.

Planning for a night of imaging

Like anything else, it's good to have a plan of action before you start imaging. Clear skies are precious, so don't waste any of that time figuring out what target you want to shoot or discovering that it is blocked by trees, too close to the horizon, or lost in light pollution from a nearby city.

Know your skies. Where is the worst light pollution? Where are the obstructions (houses, trees, mountains)? Get a general sense of this, and perhaps mark it on a star chart or planisphere. What is the phase of the moon, and is that going to get in the way tonight? There are tables in Appendix B to help determine when the moon will be in the sky and for how long. Will you have to end the imaging run early as the moon rises, or start it later once it has set?

Check the weather. Are clouds going to roll in late, or worse, is there a chance of rain? While the local weather forecast is a good start, "clear" to the weather service is not always clear for astronomy. An indispensable resource for amateur astronomers is www.cleardarksky.com. Created by Attilla Danko, it uses information from the Canadian Meteorological Center to generate a detailed prediction of not only cloud cover, but seeing and transparency as well. Even better, it accounts for the phase and position of the moon, providing a complete summary of the skies from an astronomical perspective for the next 48 hours.

Know your target. Is it big enough to show any detail at your focal length? When will it transit (cross the meridian, the imaginary line through the north pole and the zenith directly above you)? With an equatorial mount, you can only track a little past the meridian before you will have to flip the telescope to the other side and realign the object.

If your object is already past the meridian when you start, how many hours will it be viewable? Will it be too close to the horizon all night? If the object is in the east, would it be better to wait until next month (when it will rise two hours earlier), and is there something better-placed now? And consider how to position the camera to frame the scene, especially for large targets or when you are trying to capture multiple objects in one field of view. Don't get stuck in a rut with every exposure oriented where north is up.

No target is easy. While a few minutes of integration time on the brightest objects may yield something that looks recognizable, even for the showpiece objects in the sky, it seems there is always more detail or subtlety to be revealed by greater integration time. The best images of "bright" object like M31, M45, or M42 still take many hours. The best images are usually the result of multiple nights of data. There is a point of diminishing returns (you'll know it when you hit it for your skies), but there's also no point gathering only enough data to produce a mediocre image. Collect as many photons as you can.

An image capture workflow

Once you have your equipment set up and properly working, it's time to go out and take the exposures. The following is a brief workflow for an evening of imaging.

Daytime preparation

- Choose your target object and decide how it will fit in the field of view and how you will orient the camera. Be sure the view leaves some space around the object. This will not only give you more flexibility in choosing how to orient the final image, but it allows for some loss around the edges due to slight misalignments between exposures.
- Determine what duration exposure you will use. Consider whether the scene is so bright that you'll need to take two different exposure durations to capture the full dynamic range.
- Make sure everything is set up. If you are using battery power, is everything charged? Is the mount facing north and in home position? Are all of the cables connected?
- Check the power conservation settings on your computer to make sure it won't go into hibernation after a few minutes of inactivity. It's fine for the monitor to go off, but a sleeping computer can stop an imaging run.
- For DSLR imagers who also use their camera for daylight photography, make sure the camera is set for astroimaging: raw file format, correct ISO, manual mode, and bulb exposure time. (There's nothing worse than discovering the next morning that you've taken a night of perfectly focused and guided… low-resolution JPEGS.)

Evening/twilight

- Balance the mount. When balancing an equatorial mount, it is better to err on the side of having little more weight to the east. This way, the right ascension motor gears are always engaged as it moves the scope toward the west. This keeps the transfer of movement from the motors tight, since slack can lead to less consistent movements when autoguiding. After a meridian flip, keeping the weight to the east requires a different position on the counterweight bar.
- If you are using a goto mount, just after twilight is a good time to align, since you'll be using only bright stars for that anyway.
- If the scope is sufficiently cooled, you can set initial focus on one of the alignment stars. Check focus periodically, especially if it gets substantially cooler as the night goes on or when you change filters.
- Twilight is also a good time to polar align, whether you are using a computer-assisted method or drift alignment.
- Flat frames can be taken the evening before or morning after imaging. See the next section for more information on the options.

After dark

- Slew to your target object.
- Take a few test shots for composition. Adjust the framing of the scene if necessary, but be careful not to disturb the focus if you have to rotate the focuser. Fractions of a millimeter count here; don't hesitate to go back to a bright star and refocus if things don't feel right.
- Use a full-length test shot to verify that the planned subexposure time is long enough.
- If you are autoguiding (and you almost certainly should be), start your autoguiding software, pick a guide star, calibrate, and let it settle in until the guiding is stable. Once guiding is confirmed, start taking light frames, either through software or a handheld timer device.
- If you're using a DSLR, turn off the back-of-camera display after you're done focusing and aligning, since it generates additional heat.

- If you're staying awake, check on things periodically to make sure everything is stable. Have a look at the tracking graph, and maybe even go out to enjoy the stars visually with binoculars or another scope. (Yes, you can actually see things with your eyes too!). Refocus as needed. Depending on the crime level in your neighborhood and the timing of any meridian flip, you might be able to go inside and go to sleep.

The next morning

- While the camera is still in exactly the same position, take flat frames. Whatever evenly illuminated light source you use, take your flats while the imaging chain is the same as it was for the light frames. DSLR users should turn off the sensor cleaning function so that any dust motes on the sensor stay in the same place as they were when the light frames were taken.

Daytime or cloudy nights

- Stack and process the images in pursuit of aesthetic perfection.
- Take library dark and bias frames when you can't image. This can often be done indoors in a dark place.

Dark frames

The process for taking dark frames depends on whether your camera has regulated cooling. Most dedicated astronomical cameras with thermo-electric cooling (TEC) can dial-in a specific temperature. As long as this is within the cooling ability of the TEC, usually 30–60° C below ambient, a precise temperature can be maintained for all exposures. This allows you to take all light frames at one temperature. In this case getting matching dark frames is just as simple, since you can set the same temperature and take darks at any time. So aside from the benefits of lower thermal signal that come from cooling, the ability to regulate that cooling makes collecting dark frames much easier.

Those using cameras without regulated cooling, like DSLRs, have a more complex situation. There are two approaches to calibrating thermal signal for them. The first is to use software to mathematically scale a master dark frame (an average of many dark frames at one temperature with the bias signal subtracted) to approximate the thermal signal level of light frames that were taken at different temperatures. This is possible because the thermal signal is reliably linear: for nearly all sensors, a change of about 6–7° C (about 11–13° F) results in a doubling or halving of thermal signal. Software like MaximDL can use a master dark and scale it to match the temperature or exposure time of the light frames.

The second approach to dark frames for unregulated cameras is to maintain a library of them taken at many different temperatures, and then match those with light frames. This requires some effort, since you must match the temperature at the sensor as well as gain setting and exposure duration. It helps if you only use one or two gain settings and exposure times for all of your imaging, otherwise there are too many variables in play. When matching darks to a set of lights, choose darks that are as close as possible in temperature to the lights. If the temperature varied substantially while shooting the lights, break them into groups during calibration and apply different darks to each to ensure a close match.

It is crucial that the temperature *at the sensor* is used to match the dark frames with the lights. In cameras that are not actively cooled, that temperature will be higher than the ambient atmosphere. Never rely on a reading of the ambient temperature when matching dark frames. If the dark frames are taken with the camera away from the telescope, this can result in darks that are very different from lights taken on a night of the same temperature. The telescope can act as a heat sink, and holding the camera off the ground is usually cooler than a camera resting on the ground. If you took your library of dark frames with the camera resting on the ground or in a box, the sensor will be warmer than it would be when attached to a telescope on the same night.

The sensor temperature can also change when the ambient temperature is constant. The first few exposures are generally cooler as the camera warms up, then the internal temperature stabilizes. If this difference is more than a few degrees, it can lead to mismatched darks and lights, so on warmer nights, some DSLR imagers will discard the first few lights and the first few darks. Fortunately, most DSLRs record an approximate sensor temperature as part of the Exif metadata in the raw file. This information is readable by any of several available programs (like DarkLibrary and DarkMaster) which can guide you as to which library frames to use based on the EXIF information.

Regardless of your camera, err on the side of taking a lot of dark frames. As with all exposures, we reduce noise by taking the average of many exposures. While the reduction in noise is a diminishing return as you add more frames, darks are fairly easy to take. So 16 exposures will reduce noise by a factor of four, and 25 exposures by a factor of five. These would be 25% and 20% of the original levels respectively, so you may think the additional 5% nominal reduction in noise is too small to bother with. For some situations, it is. If your noise is already low, nine more exposures may not bring any visible difference. But if you are using an uncooled DSLR on a hot summer night, you may want 50 or 100 darks, because every little bit helps when the noise is that high.

When the ambient temperature is well below freezing, even DSLRs exhibit very little thermal signal. If it is really cold or your camera is actively cooled, the dark noise is mostly driven by the pixels that are most sensitive to the thermal agitation of electrons out of the silicon. Each photosite responds a little differently to the heat, and the most sensitive photosites are known as hot pixels. Dark frames will map these pixels, and you can get by with fewer frames to accomplish this when the sensor is cold. When it is even moderately warm however, the photosites of uncooled cameras will collect enough thermal signal that some of the hot pixels will saturate.

Finally, be careful to take your darks in the dark. Many mechanical shutters are not perfectly sealed, and even a few photons can contaminate your dark frames. Figure 60 shows how a small amount of light leakage can build up in a 10-minute dark frame, which was taken in the morning twilight.

Figure 60. A dark frame contaminated by light leakage along the bottom

During warm weather, it is sometimes recommended that imagers with uncooled cameras add a "cool down" period between exposures, the idea being that a gap in exposures will allow some of the heat produced by the camera's electronics to dissipate, reducing dark signal. However, it is important to weigh the value of the lost signal that you could be collecting while waiting for the sensor to cool: the slight decrease in noise from a cooler sensor probably has a smaller impact on the final SNR than if you had used the cooling time to collect more exposures. In other words, seize the opportunity to capture as many photons as possible.

Flat frames

The primary goal of flat frames is to correct for differences in illumination, so they require an evenly illuminated field of view. This can come from any of several sources, including the twilight sky, diffused artificial or natural light, or even a nearby wall.

A clear twilight sky is a readily available flat field, but make sure that it hasn't gotten dark enough that any stars are visible in your image. This can happen while the sky is still surprisingly bright to the naked eye. These stars cause bright spots in the flat which will result in dark spots or holes in your calibrated image. Twilight flats are also a race against time if you have multiple filters for which to capture flats.

You can shoot flats in the daytime as well if you use a diffuser. Carefully stretch a few layers of white t-shirt fabric over the end of the dewshield. In order to ensure even illumination of the fabric and avoid shadows or gradients, be sure to point the scope away from the bright sky. Scattered light from the shade is preferred.

You can also take flats indoors if you prefer. An evenly lit wall can work as long as the scope is perpendicular to it (so each area of the wall being imaged is equidistant). There are also electronic panels and other specially designed light boxes designed for flat frames. Homemade light boxes can use photography diffusers or translucent plastic to even out the light from a bulb. Another alternative is to use a blank white laptop screen (try an empty word processor document) held close to the objective.

Most sensors respond linearly to light across most of their dynamic range, with the exception that their response tends to flatten out as the well approaches saturation. It takes more photons to bring the well from 95% to 100% full than it does to go from 20% to 25% full. The goal is to have flats that are bright enough that read noise is not a factor, but also not so bright that the pixels are in this non-linear range of the sensor. As a rule of thumb, the top 20% of the dynamic range should be avoided. Wherever the flat frame's pixels are in the non-linear response range of the sensor, the light frame will be undercorrected.

For DSLRs set the camera to automatically determine the shutter speed (aperture priority mode, usually marked "Av" on the dial) to ensure that you get an exposure near the middle of the histogram. To prevent any influence from thermal signal, flat exposures should be short, usually under a second. Because read noise and bias level varies with gain, and bias frames are subtracted from flats in the calibration process, using bias frames generally requires that a consistent gain be used across all calibration exposures. However, most calibration software allows the use of 'flat darks' in lieu of bias frames. These are darks taken with the same exposure time as the flats.

For dedicated astronomical cameras, simply take an exposure that fills around half the well. For a 16-bit ADC, an exposure that puts most pixels in the 20–40,000 ADU range is fine. The vignetting for fast systems with large sensors can be as much as a 30% drop off from center to periphery, so ensure that both extremes remain in the linear range of the sensor. Since each filter can have its own dust, defects, or misalignments, those with monochrome cameras need to take a separate set of flat frames for each filter used.

Whatever light source you use, it is crucial that the imaging chain be in exactly the same configuration as it was (or will be) for the light frames. Even the slightest rotation of the camera will cause a mismatch between the flats and the lights. As mentioned, DSLR users should turn off any sensor cleaning function so that any dust motes remain in the same position for the flats as the lights.

> Except when taking flats, non-linearity of the sensor is not a critical issue for aesthetic imaging. Those doing scientific photometric work must end their exposures well before the photosite is full.

Bias frames

Since bias frames are the easiest to take, there is no excuse not to have a lot of them on hand. Set up everything as if you were taking dark frames, with everything light-tight and using the same gain setting as the lights, but take the shortest possible exposure. You don't want to capture any signal at all, not even thermal signal. The goal is to capture only the bias level for each pixel (the small voltage applied to the ADC during readout). The value for each photosite will vary above and below the bias level a little with each exposure. This variation is the read noise. The temperature is not critical here, since the read noise and bias level are not temperature-dependent.

Dithering light frames

While lights are obviously not calibration frames, the way you capture them can help reduce the impact of noise in the final image. If your sensor has banding or fixed pattern noise from the way the electronics manage the read out process, it can be helpful to dither. In this context dithering means to shift the telescope a few pixels between each exposure. This creates a tiny bit of offset between each subexposure that reduces the effects of bad pixels or fixed-pattern noise generated by the camera's electronics. Once you are comfortable with autoguiding, dithering can be an easy additional step to improve image quality.

Imagine a small star in the scene. With perfect guiding, the light from the star will fall on exactly the same pixel in every subexposure. By moving the telescope at random a little between each exposure, its light falls on a different pixel in each subframe. The subframes will all be aligned in calibration, but any fixed pattern noise from the sensor will be mitigated because different pixels will contribute to that star's data in the aligned image.

Nebulosity automates dithering when used in coordination with PHD for guiding. MaximDL and CCDSoft can also automate dithering when used for guiding. (CCDSoft achieves this through a free plug-in called AutoDither from Paul Kanevsky.)

12 Atmospheric effects

Light pollution

Even the darkest sites in the world aren't perfectly dark. There is always some level of natural skyglow due to ionization in the atmosphere, zodiacal light, etc. But if you live anywhere near the population centers of the world, the thought of zodiacal light contributing to your light pollution may draw a chuckle (if only!). There is no doubt that man-made light pollution makes imaging more difficult, but there are still plenty of successful imagers in urban and suburban areas, and there are steps you can take to minimize its effects.

It helps to plan ahead and maximize the best parts of your sky. Light domes from urban areas may cause severe pollution close to the horizon, but as we'll see later, the lowest altitudes are not suitable for most imaging anyway. If the area around the zenith is clear and dark, there is an opportunity for imaging there. Seek objects that will spend the most time near the zenith for a given night. The time of year also matters. Light pollution can be seasonal, varying with humidity levels and even snow cover on the ground, which reflects light, exacerbating light pollution. Some seasons may be better for imaging than others.

Even where there is little man-made light pollution, the moon is a natural source of light that overwhelms most diffuse deep-sky objects. For dim targets like diffuse nebulae, even a small amount of moonlight is too much for broadband imaging. Fortunately, the moon interferes with imaging less than half of its cycle. When the moon is waning, you can get progressively more dark imaging time in the evening before moonrise, especially past 3rd quarter. When the moon is waxing, it sets earlier, and until after 1st quarter there is usually sufficient dark time to get most of a night's imaging done before sunrise. With the shorter nights of summer, the impact is greater, while in winter, there may still be plenty of imaging time left after the moon has set or before it rises. See Appendix B for tables that show the number of moonless hours for given dates and latitudes.

Target choice is important as well. Globular and open clusters are tolerant of light pollution because of the concentration of their light into points. Diffuse nebulae are most sensitive, though emission nebulae can be captured with narrowband imaging. Dim broadband targets like reflection nebulae and galactic dust lanes require truly dark skies for the best quality data. Given a choice between the M13 globular cluster and the Witch's Head Nebula, the urban imager may want to save the famously dim nebula for a trip to a dark sky site.

Atmospheric transparency is another factor in light pollution. Particulate matter decreases transparency by its presence and also by providing nucleation sites for water vapor to condense. Light pollution is scattered by this atmospheric haze, making the sky background brighter. Specific astronomical weather forecasts predict transparency. Again, imaging near the zenith usually provides for greater transparency, as you are looking through less atmosphere.

Figure 61. Suburban light pollution is worst near the horizon

Sometimes when the transparency is poor, however, the "seeing" can be good. The stars barely twinkle, and the air is still. On these nights, resolution can be great even if the transparency is mediocre. Try imaging brighter targets where the improved resolution will help. Planetary imaging is a case where seeing is more important than transparency. Some deep-sky targets like open and globular clusters are less affected by transparency than seeing because they are the sum of point sources of light rather than diffuse objects. Even the brighter planetary nebulae are best imaged when seeing is excellent.

Filters are possibly the greatest weapon against light pollution. For a limited set of objects, narrowband imaging allows urban imagers to capture images that would be little different at darker sites. And for broadband objects, light pollution filters can reduce the effects of man-made light pollution. When choosing LP filters, note that some are

Acquiring Images

designed for visual use and others for imaging. The imaging filters are designed to produce more balanced colors than those designed for visual use only.

Target altitude

Celestial objects are difficult to image near the horizon. As you approach the horizon, the view passes through more air, which has several negative effects. More air means greater turbulence (poor seeing), greater atmospheric extinction (dimmer views), and greater refraction (shifting the position of objects). As we'll see, the last two—extinction and refraction—are wavelength-dependent, so they can alter an object's hue and spread its light out into a spectrum. Some mounts will even have trouble tracking properly since the apparent motion of objects slows near the horizon due to the increased refraction the light passing through more atmosphere. Close to the horizon, this refraction causes objects to appear higher than they really are. At the horizon, the effect is about half a degree—the width of the full moon. The software in most computerized mounts can correct for this effect with varying degrees of accuracy.

The atmosphere's density decreases by half for approximately every 5 km increase in altitude, thus about 90% of the atmosphere is within 15 km of sea level. Looking straight upward to the zenith is the view that passes through the least atmosphere. All else equal, imaging objects near the zenith will produce the clearest images, because there is less air between you and your target. This shortest path is referred to as one airmass. As you look down toward the horizon, the view passes through more of the atmosphere, and the more air you look through, the greater the effects of turbulence.

An estimate of how many airmasses you are looking through for a given angle is determined by

$$a = \sec(90° - h)$$

where a is the number of airmasses and h is the angle from the horizon. This formula is accurate down to about 15° from the horizon, and then it starts to break down due to factors like curvature of the earth, refraction, and variations in the atmospheric density and temperature. There are more complex formulas, but you are unlikely to need them unless you are doing photometry. At about 30° above the horizon, you are looking through two airmasses, which is where imaging can starts to be a challenge. Anything below 15° is usually too difficult to bother with.

Figure 63 depicts the number of airmasses at 15 degree increments. Note that as you go below 15° toward the horizon, things get ugly fast. If you are looking through about 15 km of atmosphere straight up to the zenith, the view at the horizon is looking through over 500 km!

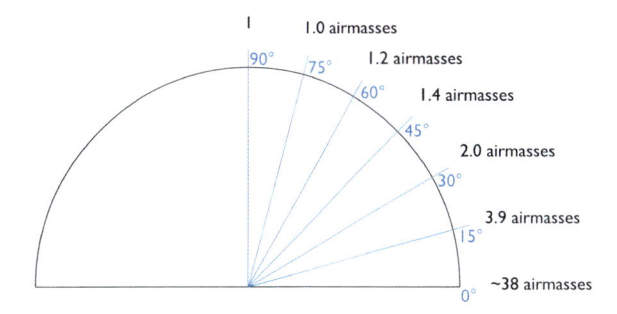

Figure 63. Airmass by angle

Atmospheric extinction causes objects to look dimmer by scattering their light. This effect is dependent on the wavelength of the light. The blue end of the spectrum is more affected by extinction (which is why sunsets are orange-red), so when planning filtered imaging runs, it is best to capture blue or OIII exposures closer to the zenith, moving toward longer wavelength filters closer to the horizon. This

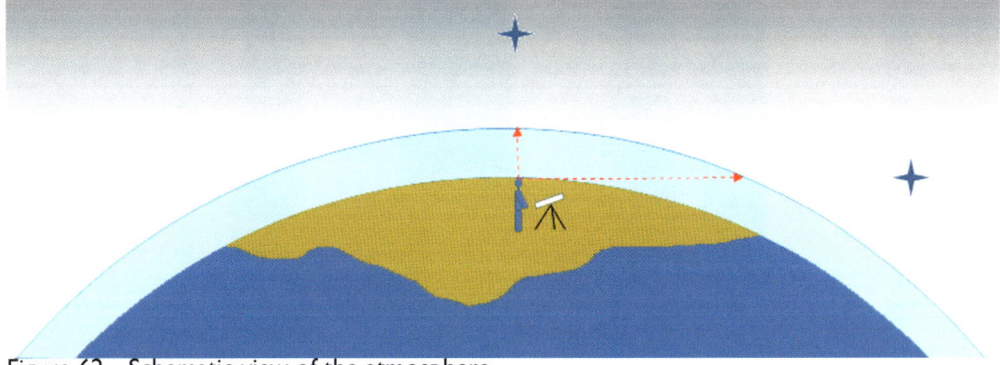

Figure 62. Schematic view of the atmosphere

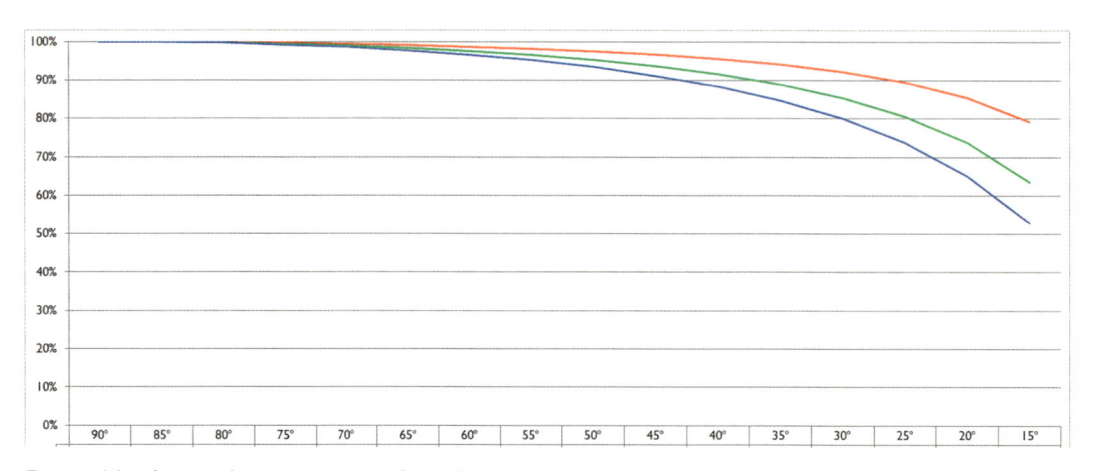

Figure 64. Atmospheric extinction by color

differential effect on colors will also shift the hue of an object toward the red end of the spectrum. The approximate effect of atmospheric extinction on the brightness of the primary RGB colors is shown in Figure 64.

Extinction is not the only altitude-related effect on the view. The atmosphere behaves as a lens with a very low refractive index. Like any lens, it refracts different wavelengths of light different amounts. The index of refraction for visible light through dry air at 15° C and standard pressure ranges from 1.000276 at 700 nm to 1.000283 at 400 nm. By comparison, most glass has a index of refraction between 1.4 and 1.8. However, we look through several millimeters of glass and several kilometers of air.

Total refraction increases with decreasing angles relative to the horizon as the view passes through more air. Due to this differential effect, the red end of the spectrum is refracted lower than the blue end, smearing the image out into a spectrum much like a prism. The result can look similar to chromatic aberration, with bright stars showing a blue fringe on top and red fringe on the bottom.

Figure 65 plots the apparent differential refraction in arcseconds relative to 550 nm light for angles down to 20°. The differential refraction at low altitudes can be several arcseconds between the red and blue ends of the spectrum, and the effect is greater on short wavelengths of light. (These values are approximations for 15° C dry air at sea level using the formulae presented in W. M. Smart's 1931 *Textbook on Spherical Astronomy*. Greater humidity, altitude, and temperature reduce the refraction index of air, but the plotted results illustrate the general effect.)

Planetary imagers overcome differential atmospheric refraction with optical devices known as atmospheric dispersion correctors. These devices use two wedge-shaped

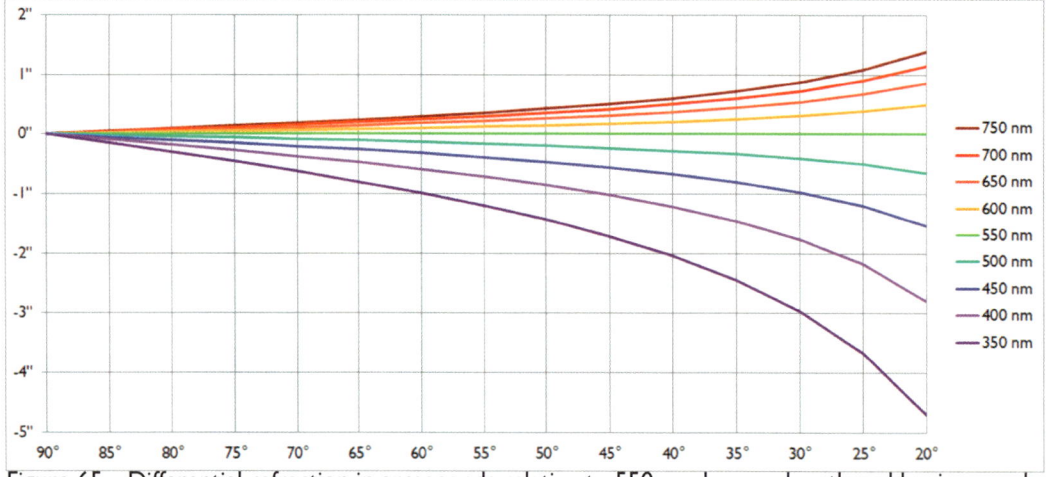

Figure 65. Differential refraction in arcseconds relative to 550 nm, by wavelength and horizon angle, assuming dry air at sea level

prisms with an adjustable offset between them to vary the total angle of the prism. Using RGB filters can also mitigate the problem. Since each filter passes only a third of the visible spectrum, the apparent angular separation in each image reduced to a third of the full spectrum effect.

> The rare "green flash" above a rising or setting sun is a result of the extreme refraction at low horizon angles. As with light shone through a prism, the blue/green end of the spectrum is refracted more than the red, so those colors of light remain visible as they are refracted from below the horizon after the red-orange light from the top of the sun has set.

Local turbulence

Atmospheric turbulence from nearby sources also plays a role in final image quality. While the overall atmosphere is out of our hands, the conditions of the air near our observing site might be. If possible, avoid views that pass over nearby asphalt, roofs, or any other large heat-absorbing space. The difference in temperature after nightfall will lead to turbulent air until the object equilibrates with the air temperature. Being downwind of these areas can present a problem too. Even houses can cause significant turbulence—on cold nights the heat from a chimney can leave a trail of turbulence across the sky.

More importantly, your equipment must be equilibrated to the ambient temperature. Not only should the telescope itself cool off to prevent tube currents and ensure stable focus, but a tripod or mount that got hot in the evening sun can take hours to cool down as well.

13 Diagnosing problems and improving image quality

There are a lot of things that can go wrong with images before we even begin to calibrate or process them. These dara are the raw material for what are hopefully beautiful deep-sky images, so the old maxim of "garbage in, garbage out" is particularly appropriate. There is little we can do at the computer to correct for fundamental problems in the original exposures. If there are problems with some exposures, the best strategy is to exclude them from the calibration process entirely.

We've already seen the effects of poor focus and inaccurate polar alignment. This chapter lists some other common imaging bugaboos, as well as some simple tips to help improve image quality.

Wind or tracking errors

Wind gusts and glitches in tracking can ruin a subexposure, creating smeared or doubled stars. Especially when taking long exposures, this can cost you a lot of lost data. Figure 66 shows an exposure of the Rosette Nebula that was ruined by wind gusts.

Figure 66. Wind gusts ruin exposures

Similarly, Figure 67 shows a close crop of a 20-minute exposure where the mount wasn't balanced—the counterweight was loaded too far to the west. Because the gears were not consistently engaged as the mount turned to the west, it wobbled a little in right ascension. The result is that the streaks all run the same direction throughout the image along the angle of the right ascension's axis. (Hot pixels are also visible in this uncalibrated image.)

Figure 67. Gear slack is seen when the mount is improperly balanced

Mirror flop

Most Schmidt-Cassegrain and Maksutov-Cassegrain telescopes focus by moving the primary mirror. As the mount moves the telescope across the sky, the heavy mirror can shift as gravity pulls on it from a different side, causing a change in focus. Models designed for imaging will feature a mirror lock.

Diffraction patterns

Any sort of obstruction in an optical train will cause a diffraction pattern in the focused image. The vanes in most Newtonian and Ritchey-Chrétien reflectors produce a characteristic diffraction spike that many find aesthetically pleasing, even if it acts as a mild point spread function across the image. A perfectly focused refractor should produce round stars. The Airy disk is not usually visible since imaging is rarely done at the focal length and resolution required to reveal it.

Another diffraction pattern that sometimes appears in images taken with refractors is a sort of "shadow-spike" that is usually the result of spacers between the lens elements. The spacers are not always visible looking directly at the objective, but their result in the final image is clear. An example is shown in Figure 68. Unlike the spikes from reflectors, they are generally considered to be a minor optical defect, though something many imagers can live with as long as the effect is contained to the brightest stars.

Figure 68. Diffraction shadows from lens spacers

Many CCD sensors now incorporate microlenses into the photosites to direct more light onto the silicon, especially when anti-blooming gates are present. Because of their square shape, these microlenses can create diffraction spikes of their own. These spikes will always align with the axes of the sensor, not the telescope.

Figure 69. Microlens diffraction spikes

Focus

Problems achieving focus can be the result of astigmatism in the optics, inadequate color correction, non-orthogonality between the sensor and lenses, or it could just be something as simple as dew on the lens or mirror. Poorly corrected field curvature can prevent simultaneous focus at the image center and the edges. And remember that focus can shift either through mechanical slippage or temperature changes.

Any telescope with refractive elements can have bloated stars from unfocused UV and IR light. Cutoff filters can eliminate this bloat. Since CCDs and CMOS sensors are sensitive to a wider range of wavelengths than the human eye, even a telescope that produces no chromatic aberration visually can show bloated stars in long-exposure images.

Halos

Halos around stars, especially concentric rings, are usually the result of internal reflections between glass filters or coverslips. Light that passes through all the surfaces arrives in focus at the sensor, but light that bounces between optical surfaces on its way to the sensor travels further than the focal distance, so it arrives out of focus, creating a halo.

Halos are most often seen with CCD cameras with sensors sealed under glass for cooling and when using filters. If you see halos in the final image, the first thing to check is if they occur with only one filter or with all of them. Also check the view through an empty filter slot in the wheel. If there are several potential reflective surfaces near the sensor, the size of the halo can be used to help diagnose the problem. The diameter of the halo is proportional to the distance between reflection surfaces, and faster focal ratios will produce larger halos than slower optics for the same distance. The extra distance light must have traveled (beyond the focused distance) to create the halo using the formula:

$$Distance\ traveled = \\ halo\ size\ at\ sensor \times focal\ ratio$$

Since the light must have changed directions twice to end up back on the sensor, the actual spacing between surfaces is half of the distance traveled. The extra distance traveled is not necessarily the distance of the offending surface from the sensor, since reflections often occur between surfaces, even between the front and back surfaces of a filter. Draw out the distances between all surfaces in your system to see if there is an interval that matches the halo size. Note that any portion of the distance traveled through glass should be divided by its refractive index (1.5 is a reasonable estimate for most glasses).

It is hard to avoid some haloing around the brightest stars—second magnitude Alnitak is a perennial bugaboo for those imaging the nearby Horsehead and Flame Nebulae, as shown in Figure 70. The quality of anti-reflective coatings also varies from brand to brand and filter to filter.

Figure 70. A halo around Alnitak

Exercises

2.3 The primary reflection halo around Alnitak in **Figure 70** is 110 pixels across, and the imaging camera had pixels that were 5.4 μm square. The telescope was f/5.3. How far from the sensor was the offending reflective surface?

Tips for better image capture

These are a few major steps forward along the astroimaging learning curve, each leading to better image quality.

1. **Improve your signal-to-noise ratio, through darker skies, filters, or longer total exposure time.** SNR is the ultimate factor in determining image quality, and you can improve it by increasing your signal (longer total exposure time, faster optics) or decreasing the noise (darker skies). Higher SNR allows the dimmest areas of an image to withstand the stretching needed to reveal their detail. While truly dark skies are the benchmark, traveling to them is not an option for everyone. Light pollution filters can help reduce skyglow for normal color imaging, and narrowband filters nearly eliminates its effect on emission objects. Regardless of your skyglow, more data is always better than less. Dedicating more time to a target, usually over several nights, will yield the best results. Think of deep-sky images as projects, not snapshots.

2. **Take sufficient calibration frames.** This is the easiest technique to improve your images. Especially for uncooled cameras, think 50 darks and flats, not five. Bias frames are so easy, there is no excuse not to take plenty. Why increase the noise in your images because of insufficient calibration frames?

3. **Attain and maintain optimal focus.** Don't settle for close-enough focus. Use a focus mask or electronic focusing. Refocus periodically, especially when the temperature changes. Your final images will reward you with pinpoint stars and fine detail.

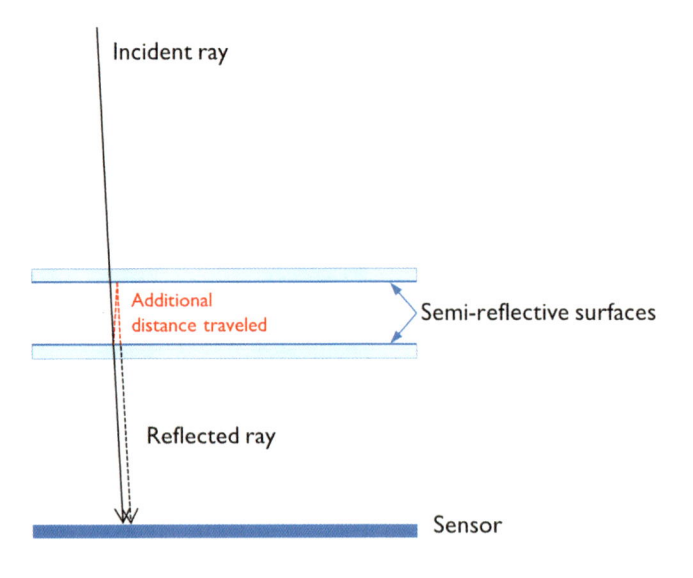

Figure 71. Reflective causes of a halo

4. **Autoguide.** For less money than a decent eyepiece, you can have a computer automatically guide your mount with greater accuracy than any human could achieve. That means less wasted exposure time, tighter stars, and better resolution. Once you have it set up, it's almost effortless.

5. **Shorten your setup time.** If it takes an hour to set up, you are not as likely to image as you are if you have it down to a 10 minute process. Not everyone can have an observatory, but there are small steps that can make your life easier. Tie together cables and electronics. Mark a roughly aligned spot for tripod legs. Consider a pier. If the thought of lugging a 300 mm reflector onto the mount means you stay on the couch, get smaller optics. As with visual astronomy, the best telescope is the one you'll use most.

6. **Gain practice and experience with image processing.** Processing images is as much of an art as a science, so there is no substitute for practice. Stick with one program and become familiar with all its options and tools. See how others process their images, but learn the underlying principles of image processing so you can make your own decisions. The next part will introduce these concepts and show examples of the processing techniques that translate great data into beautiful images.

3 Processing Images

Once we've collected quality image data at the telescope, it's time to go to the computer to bring together all of the exposures into one beautiful deep-sky image. At a high level, there are three major stages of work needed to turn raw data into a pleasing image. First, the individual subframes need to be "stacked." This means carefully aligning them, then averaging them together while correcting for the imperfections of the sensor and optical system. The resulting image is then "stretched," re-mapping the image across a wider dynamic range, emphasizing specific features. Finally, more subtle adjustments are applied selectively to create a pleasing final image.

This process can be a source of frustration due to the overwhelming number of tools and options available. We'll cover the underlying principles and software tools used to calibrate and process images. And since nothing explains better than example, there are detailed examples of image processing techniques.

NGC 6960 and Pickering's Triangle in Cygnus, taken with narrowband filters
(9 10-minute exposures through H-alpha, SII, and OIII filters)

14 Color in digital images

Before we start processing images, we need to take a brief detour to understand the image files we work with and a little bit of color theory. While we take it for granted that an image will display properly when we open it, there is a lot going on behind the scenes that can affect image quality and color accuracy.

File formats

It is critical in astroimaging to work with raw data from the sensor. Dedicated astronomical cameras are designed with this in mind, applying little or no correction by the camera. Most DSLRs can also do this with the proper settings. For normal daytime photography, most users are happy to have the camera automatically adjust their images to improve color, contrast, and correct for minor defects. But for astronomical imaging, we need the camera to keeps its hands off our data for two main reasons. First, because we are going to calibrate the exposures ourselves, correcting for the optical system and sensor characteristics. And second, because the dynamic range of raw astronomical images is nothing like that of daytime images.

Like any consumer camera, DSLRs can record data in JPEG (Joint Photographic Experts Group) format, which is a standard for internet images because of its small file sizes. This format, however, uses "lossy" compression that does not retain all of the original image data to reduce the file size. Further, JPEGs are usually processed in the camera to alter color, contrast, and other aspects of the image in a way that produces pleasing snapshots. Worse, the dynamic range of JPEG files is substantially reduced. If you are using a DSLR for astronomical imaging, you must shoot raw. After final processing, you may want to create a JPEG version of your final image for web display, but this format should never be used for intermediate steps.

Dedicated astronomical cameras typically produce either TIFF (Tagged Image File Format) or FITS (Flexible Image Transport System) image files. These can be grayscale or color, with at least 16 bits per channel preferred. In addition to the actual image, all of these file formats have the capacity to store metadata about the image.

TIFF is a flexible file format that acts as a container for image data and a header containing metadata about the image, thumbnails, or other information. Nearly any bit depth or color space is possible within a TIFF file. While it is possible for a TIFF to hold data in formats such as JPEG, in practice they almost always contain image data in a lossless compression format that preserves the underlying data. TIFF is the preferred file format for graphic arts work or anytime high quality image data is needed.

FITS is a file format that was specifically designed for astronomical and scientific imaging. It is also highly flexible through the use of a metadata header that describes the subsequent data, so the actual data format can be vary, but is typically lossless TIFF. The header of a FITS file is human-readable, with user-defined keywords allowed. The format was designed by professional astronomers in the early 1980s for use on magnetic tape media, but its flexibility has allowed it to remain the dominant format for decades. FITS files can become very complex due to this flexibility, but most imaging programs stick to a basic version that is widely interpretable.

The raw files recorded by DSLRs are usually proprietary formats that vary by manufacturer. Canon's cameras produce .CR2 or .CRW raw files, Nikons produce .NEF, and Pentax cameras can output .PEF or .DNG (a cross-platform raw format). All of these can be opened by imaging software with the appropriate plug-ins or updates. Like TIFFs, the raw image formats used by DSLRs can also add metadata about the image in a structured header. This data uses a standard known as Exif (Exchangeable image file format) that typically captures information about the camera and lens used, exposure details, date, time, and most usefully, camera temperature.

Nikon's raw files stand out in that they are not always truly raw—some processing is applied even to supposedly raw files, and this degrades their usefulness for astroimaging purposes. This processing is designed to reduce noise, but it can interpret dim stars as hot pixels and eliminate them.

Visual response to color

Our eyes are tuned to the spectrum of sunlight passed through our atmosphere, and we perceive light with that balance of brightness across the spectrum to be white. We take our color vision for granted, but our brain plays a large role in color perception, using context and expectation to establish what color we 'see.' Our evaluations of brightness and color are highly subjective. Electronic sensors account for none of these external effects, so it helps to understand a little more about how the experience of color is created.

Animals with a single type of light sensitive cell see only monochromatic levels of brightness. Most mammals have two types of cells, each responding to a different range of wavelengths, which allows limited color differentiation. Some animals like birds, fish, and insects can have visual systems using four or even five types of color sensing cells with sensitivities that extend into the ultraviolet region of the spectrum.

Humans have three different types of color sensitive cells, known as cone cells, on our retinas. These have peak sensitivities of approximately 570 nm, 540 nm, and 430 nm. Based on the relative outputs of these cells, our brains create the perception of color. This perception is not precisely linked to the wavelengths of light that enter our eyes, however. Consider that your TV or computer monitor can only emit light in three discrete ranges that appear to us as red, green, and blue. It cannot emit light of the wavelength we call yellow (around 575 nm). In order to create a perception of yellow in your brain, it simultaneously emits red and green. This red+green combination stimulates the three types of cone cells in the same ratio that true yellow light, triggering the perception of yellow.

Colors are creations of the mind, a way of helping us distinguish objects in our surroundings, so their perception is inherently contextual. For instance, brown is not a specific part of the spectrum of emitted light (can you find it in a rainbow?); it is the way we perceive dim or unsaturated yellow in the presence of brighter surroundings. Yet most tones of blue and green light are still perceived as blue or green whether they are light or dark.

Our brains are also adept at correcting our perception of color even when the ambient light is not white. Looking at a white piece of paper under fluorescent lighting reflects very different wavelengths as when the same paper is viewed by candle light, but you see it as white in both situations. Further complicating things, our mental color palette is not directly related to the wavelength of light. This is why scenes shot indoors under incandescent lights appear yellow when shot using your camera's outdoor color balance setting, even though you don't perceive a white wall as yellowed under the same circumstances.

Producing color on print and screen

Even though electronic sensors faithfully record the photons they capture, a color image file is interpreted differently by an electronic display or printer. The same image will look different on different monitors or even when using different video cards. A printed image will look very different than the same image displayed on a monitor, since light reflected from pigments covers a different range, or gamut, than a monitor's emitted light. Printers using different pigments or dyes will produce prints that look different. Even the same print viewed under sunlight will look different when it is viewed under incandescent lighting.

Imaging sensors are quite different from the human eye. They are receptive to a wider range of wavelengths, and their relative sensitivity across the visible spectrum is different. Like an eye without a brain, electronic sensors can only produce color images with additional processing.

In order to ensure perceptual color matching across different devices and media, there is a process of interpretation going on, mostly behind the scenes, known as color management. Color management consists of mapping the input values coming from an image file and the corresponding output (as a human would see it) from a particular display or printer. This creates a profile for each device that tells the computer what to expect and how to adjust the inputs to create a desired output.

More broadly, color spaces are descriptions of how to encode colors into numbers that can be stored in image files. Color spaces may be specific to a device or device-independent. The transformation of values from one color space to another is what maintains visually consistent results across settings. All of this work is done by device drivers, operating systems, and imaging software. But to ensure that your images are correctly displayed, it helps to know a little about color management and how colors are reproduced.

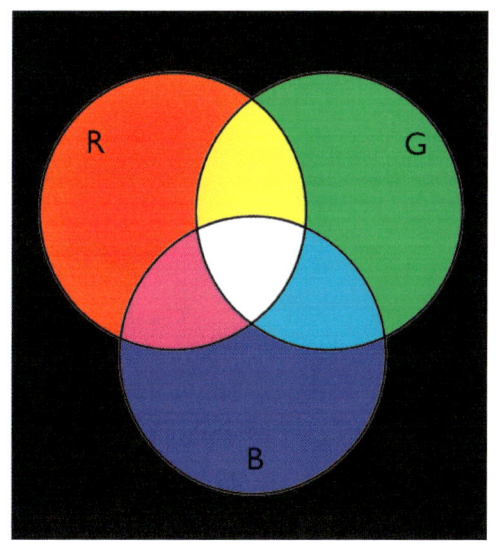

Figure 73. Additive (RGB) colors of emitted light

Image files usually contains values for the intensity levels of red, green, and blue for each pixel, which is how colors are created by emitted light devices like monitors. This is known as the RGB system. Figure 73 shows how the three primary emitted light colors mix together. No color is equal to black, since this is emitted light. All three colors combine to make white. Each combination of primary colors makes a secondary color, which in the RGB system are cyan, magenta, and yellow.

The RGB scheme for defining colors, however, is only one of many ways to describe color in an image file. Paints and inks reflect specific colors of light, absorbing other parts of the spectrum, so combining them is a subtractive process. This requires a different set of fundamental colors to represent the visible spectrum. It turns out that the best set of ink colors for printed materials is cyan, magenta, and yellow—the secondary colors in the RGB system. Figure 74 shows how these primary subtractive colors mix.

Here, the absence of pigment is white, since we normally print on white paper. The secondary colors of the CMY system are red, green, and blue. Black is usually added as a fourth to CMY, since combining all three pigments doesn't make a very dark black. This system is known as CMYK (K is for "Key" or "blacK"), and it is how nearly all color printing is done.

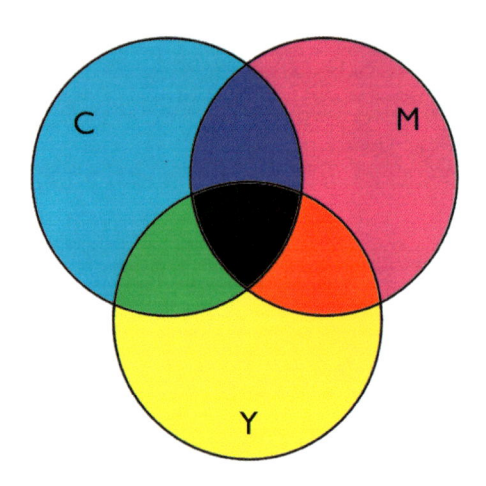

Figure 74. Subtractive (CMY) colors of reflected pigments or dyes

All of our image processing will be done in RGB, since we are viewing our images on a computer monitor. When it is time to print, software usually manages the process of ensuring that our work is rendered with as much fidelity as possible by CMYK devices. But cyan, magenta, and yellow are also crucial as the secondary colors when working in RGB. Cyan is the opposite of red, so if we want to decrease the level of red in an image, one way is to increase the cyan. Photoshop's Selective Color is one example of a software tool that makes effective use of these opposing colors for image processing. Figure 75 shows a color wheel with the 8-bit RGB coordinates of each color shown within its circle (red is the top value, then green, then blue).

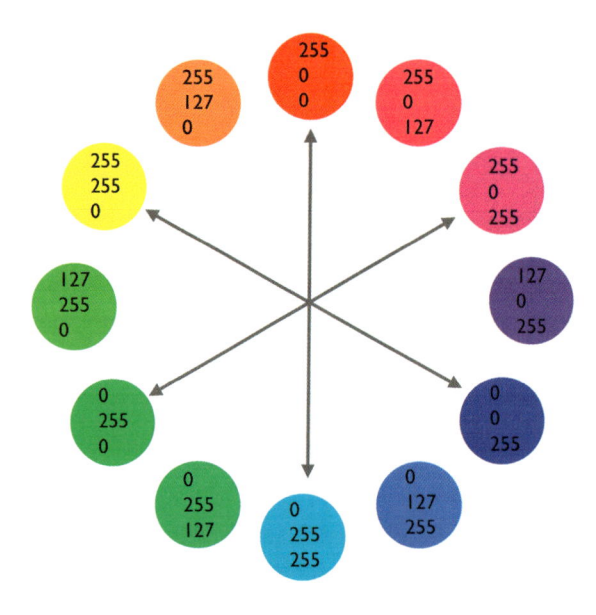

Figure 75. Color wheel with RGB coordinates

From Figure 76 and Figure 77, it is clear that the RGB and CMYK systems are opposites—the primary colors in one are the secondary colors in the other. Orange and purple, two of the secondary colors in the red-yellow-blue system used by artists, are tertiary in both RGB and CMY.

> Working in RGB or CMY takes a little adjustment if you learned in art class that the primary colors are red, yellow, and blue—combining to create secondary colors of orange, green, and purple. This is known as the RYB system. These are perfectly valid choices for primary colors, and they work well for the pigments in most paint, but using CMY allows for a wider range of colors to be produced on printed media.

Processing Images

a particular video card to drive a particular monitor, based a profile that defines how the phosphors or LCD elements in that monitor are expected to render color. When color accuracy is crucial, you can create a more precise profile of your monitor using a colorimeter calibration tool. (Popular brands include the Spyder and Eye-One.) Using a profile like this, you can be sure that what you see on your monitor is as accurate as possible.

Through color management, print devices can translate an RGB file's values into CMYK for the closest ink or pigment approximation. The translation is always imperfect because the gamuts of printed color spaces do not line up precisely with monitor color spaces. Generally, CMYK can produce some yellows that RGB cannot, and the RGB spaces contain some blues and pastels that are difficult to produce in CMYK.

Unless you are a professional printer and have specific reasons for doing otherwise, all processing work should be done in RGB, with excursions into other color models only when needed. The two most common device-independent ("working") RGB color spaces used today are Adobe RGB and sRGB. Of the two, Adobe RGB has the wider gamut of colors; it can encode all of the colors that sRGB can plus more that are outside of sRGB's gamut. Using a color space with a wider gamut is generally preferable for processing, since you can always save the final image into the smaller space.

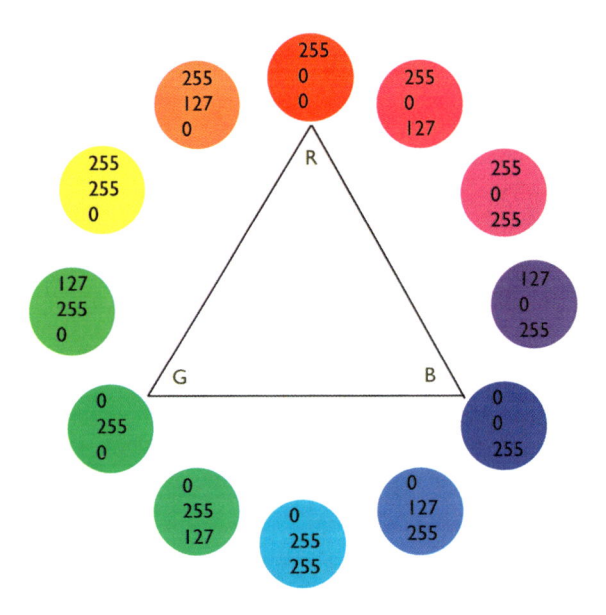

Figure 76. Red, green, and blue

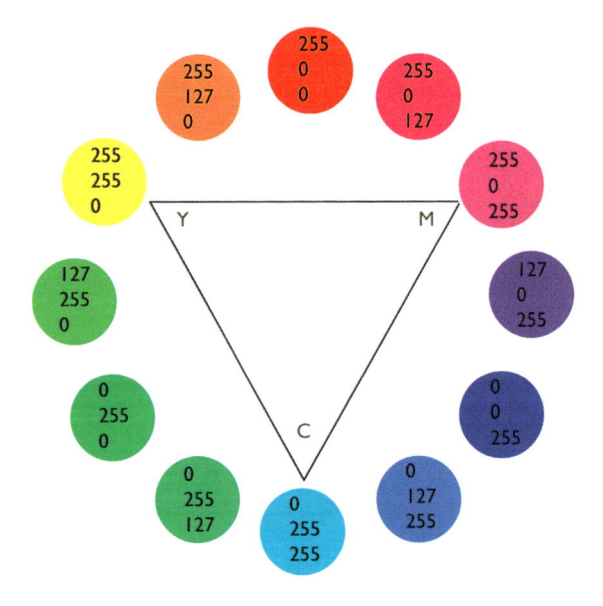

Figure 77. Cyan, magenta, and yellow

> One-shot color cameras produce raw files that are unique to a specific model, so any raw conversion software (which is sometimes part of your calibration software) must take into account the camera used. Each camera model has a unique sensor with a defined spectral response curve, specific electronics that define how the signal is amplified, and a filter array that passes certain wavelengths of light. The metadata of the raw file contains information about the camera, so most conversion software can use that information to accurately translate the raw data into appropriate color values without any intervention from the user.

Color management and color spaces

As we've seen, simply providing an RGB value for each pixel in an image file isn't enough. Since a given RGB value may be rendered differently by different hardware, the file must also designate a color space to provide context for the values. The color space defines the range of possible colors—the gamut—described by the numbers. The most commonly used color spaces are based on the way colors are represented on a monitor. The file is then translated by

LAB color

The LAB colorspace (also referred to with varying degrees of specificity as Lab, L*a*b*, or CIELAB) is a somewhat unique color space that is found as an option in only a few imaging programs, though it works behind the scenes in many. The L in LAB stands for Luminance or Lightness. This channel contains no color information, only brightness levels. The A and B channels encode the color information, with each representing one axis in a theoretical two-dimensional color space. Channel A represents a color's value on the green-magenta axis, and B represents its value on the blue-yellow axis. Together, these axes map out a space that covers the entire range of human color vision and even beyond into "imaginary" colors that the human eye does not perceive. (If you're wondering what an imaginary color is, it stems from the fact that you cannot stimulate a single type of cone cell in the human eye without also generating some signal in at least one of the other two types. Imaginary colors would be the result if you could stimulate cone cell types individually.)

Using the LAB color space for some intermediate processing steps can be very useful, accomplishing in only a few steps what would require many in RGB. Specifically, LAB is great for making adjustments to either luminance or chrominance separately. For instance, sharpening usually produces the best results when applied only to the luminance in an image. It is possible to create a pseudo-luminance layer in RGB mode, but switching to LAB mode immediately provides a pure luminosity channel to adjust. Working with the A and B chrominance channels also provides a very effective way to increase color saturation without altering luminosity. These methods are reviewed later.

Because the LAB color space contains the widest range of possible colors of any space, care must be taken when moving back to an RGB space that the colors created in LAB still map to an equivalent in RGB. Manipulations that yield imaginary colors do not predictably convert to RGB.

The HSL/HSV/HSB color model

Comparing two RGB values, it is usually far from obvious whether they are different shades of the same hue. To make things more intuitive, most graphics programs use a color picker dialog that helps facilitate the choice of appropriate colors by laying out the visual spectrum on one axis with brightness on another. These color pickers are a flat representation of a color model known as Hue-Saturation-Lightness, or HSL. Variations substitute Value or Brightness for Lightness, yielding HSV and HSB.

The HSL model is represented as a three-dimensional cylinder or cone where the hues arranged axially, with red at 0/360°, green at 120°, and blue at 240°. Naturally, the CMY colors are between the RGB locations, with yellow at 60°, cyan at 180°, and magenta at 300°. The outside edges of the cylinder have the most saturated colors, while the core of the cylinder is neutral. The third axis, from top to bottom, is brightness.

While the HSL model helps a user translate the human perception of color to a set of appropriate RGB values, it is not a color space. Any HSL-type model is inherently linked to a specific RGB color space, and translations from different RGB spaces will result in different HSL values. HSL models are also not perceptually uniform, since they are designed to linearly represent RGB values, not human perceptions. While the hue parameter is usually accurate, the saturation and lightness values can diverge widely from perception, so the model's use outside of color pickers is rare.

DSLR white balance

For normal daytime photography, the goal is to produce an image that reflects approximately what the eye would have perceived. Because the brain adjusts perception based on the hue of ambient light, we have to apply color balance to our images to accomplish the same.

The white balance setting on a DSLR is designed to mimic the "color temperature" (see box) of different kinds of lighting. This is accomplished by scaling each of the color channels by a specific amount, emphasizing the hues necessary to correct for the ambient lighting. This adjustment is only applied in-camera to JPEG or TIFF output files. Raw files are not affected by these settings other than a notation in the file's metadata about the camera settings used for the exposure. Some image processing software can read the Exif metadata and apply the white balance settings noted there, but the raw image data will still faithfully represent the unadjusted light levels received by the sensor. Since individual color channels can be adjusted in post-processing, applying a white balance setting is not necessary, though some DSLR imagers find it can save a processing step. Manual white balance settings are also used to allow a DSLR that has been modified by removing the IR filter to be used for daytime photography as well.

> The hue of ambient lighting is commonly described by its black-body radiation temperature in degrees Kelvin. A black body is an idealized opaque, non-reflective object. Any such object will emit electromagnetic radiation centered on a specific wavelength at a given temperature, regardless of its composition. At low temperatures, black-body radiation is in the infrared, but as the temperature rises, the wavelength shortens. Because of the shape of the curve, the glow progress through red, then orange, then yellow and white as it gets hotter, and finally pale blue. Only these colors are possible—there is no green black-body radiation, which is why there are no green stars. These hues are good approximations for the spectra of most natural and artificial illumination. As examples, candlelight is around 2000 K, incandescent light is around 3000 K, and overhead daylight is between 5000 and 6500 K.

Processing Images

Deep-sky color accuracy

Looking through a telescope eyepiece, only the brightest deep-sky objects show even a hint of color. This is because our eyes can integrate only a second or so of the view, and the luminance-sensing rod cells are far more sensitive than the cone cells that see color. That does not mean that deep-sky objects are truly gray. Long-exposures reveal color, though the degree to which this is exaggerated is a processing choice.

Any attempt to represent color in deep-sky objects is a subjective choice. This is true of daylight photography as well, but the differences are more acutely visible to anyone who has ever looked at a faint fuzzy through an eyepiece. We increase color saturation to make our images more pleasing, but with a daytime image, we have a mental reference. Without such a reference, deep-sky images create the temptation to over-saturate. On the other hand, without some color push, we're left with only a faint fuzzy.

Astronomical imagers also have to deal with near-IR and near-UV wavelengths, which are registered by electronic sensors, but are invisible to the human eye. Depending on the filters being used, these 'colors' may need to be squeezed into the final image's palette. Even within the colors visible to our eyes, many objects in the sky do not produce the broadband light to which we are accustomed. An entire category of astronomical objects, known as emission nebulae, emit light at only a few specific wavelengths. Normal color images of these objects are primarily a monochromatic red, but by mapping the dominant emission wavelengths to separate colors, we can create false-color narrowband images that reveal distinctions in the structure and composition of nebulae driven by different ionized gases. They can also be spectacularly beautiful.

In addition to color, photographers choose to emphasize particular regions of an image by creating greater contrast than was present in the original scene. Without dodging and burning, Irving Penn's portraits and Ansel Adams landscapes would sit flat on the page. The images they captured on film were only the beginning—the real magic happened when they carefully printed the negative. Selective contrast adjustments allowed them to create something more beautiful than the original scene, yet still faithful to it. Our aim in processing astronomical images is no different. While we should strive to be reasonable with our manipulations, like any photograph, the goal is to create a beautiful interpretation of reality.

Color calibration with G2V stars

There are some reference objects in the sky that can help establish a baseline for color balance. Recall that the human visual system is based around the specific spectrum of our sun's light, which we perceive as white. Thus we should also perceive stars with similar spectra as white. Our sun is a spectral type G2V star, so other G2V stars can act as a stellar color card. We can use them to determine the proper balance between the red, green, and blue channels in our images. This is true for most one-shot color cameras as well as RGB filtered monochrome exposures. (Since narrowband imaging is false color anyway, there is no "correct" ratio of color channels, and unmodified DSLRs can simply use a daylight white balance setting to get accurate color.)

The process for determining the correct ratios is straightforward, and it's only necessary to do it one time for a given combination of sensor and filters. First, find a G2V star that is near the zenith. Any online or software object database can help you find such a star. Since atmospheric extinction varies by color, using a star higher than about 60° will ensure that your measurements are within about 2% of the correct ratio. Focus is not critical; in fact, a slightly out of focus image will make things easier.

Take several exposures using the same duration for each exposure and for each filter. The G2V star must not be saturated in any color channel. Stack and calibrate the images as usual (this process is covered later if you're not familiar with it). If you're using color filters, calibration must be done separately for each color, including separate flat frames. Once you have the final calibrated image(s), zoom in on the star and select only the core pixels, avoiding the star's edges, which could contain color fringing from optical defects. The average brightness of these pixels in each color channel can be used to calculate the relative brightness ratios between the channels. Choose the brightest channel as 1.0 and calculate a correction factor for the other two channels. For instance if the red channel is at 40,000 ADU, the green is at 36,000, and the blue is at 30,000, the correction factors would be 1.00 (40,000/40,000), 1.11 (40,000/36,000), and 1.33 (40,000/30,000) respectively.

This ratio can be used to create accurate color images. If you are using colored filters with a monochrome sensor, you can use the ratios to determine the exposure duration. For instance, in the above example, if the red exposures were 10 minutes, the green exposures would be about 11 minutes, and the blue exposures about 13 minutes. Of course, you'll also need matching dark frames for each exposure duration.

True color fidelity can be a difficult challenge. Even if the calibrated image has perfect color balance, this accuracy can be thrown off in subsequent processing. In calibration, color channel alignment or flat frame adjustments can alter the colors. In post-processing, adjusting the black point or stretching the channels separately will also undermine the initial color accuracy. Another issue with this method is that atmospheric extinction varies slightly with color, so color ratios have to be adjusted based on the altitude of the object being imaged and the altitude of the star used for calibration.

A simpler way to use a G2V star for color accuracy is to identify such a star in your image, and then adjust the color ratios so the star is neutral. The star must not be saturated in any of the color channels, but once you've identified an appropriate G2V star, your image has a built in color calibrator. While it may sound infeasible to identify the spectral class of every star in your image to find the right one, there is an amazing piece of free software that can do just that. eXcalibrator, by Bob Franke and Neil Fleming at http://bf-astro.com, will identify the truly white stars in the field of view and calculate the appropriate levels for each color channel. You must first plate solve your image (find its scale and location in the sky), but this can be done by many programs and online services. Because eXcalibrator can usually find multiple white stars, it can provide accurate adjustments, plus the net effect of the atmosphere is inherently accounted for since the calibration stars and the object are in the same field. The program is thoroughly documented and straightforward to use. For those seeking true color fidelity, this may be as close as it gets.

The sun's G2V classification contains a lot of information about what kind of star it is. Stars are divided into spectral classes that range from hot and blue class O stars to cool and red M stars. For historical reasons, the main stellar classifications from hottest to coolest are O, B, A, F, G, K, and M (with several additional classes used for cooler stars and rare types).

The sun is a spectral class G star, which means that its light is predominantly yellow, the result of its 5780 K surface temperature. The 2 gives more specificity on where it lies in the range of class G stars by using a number from 0–9, with 0 used for the hottest within the range. Thus, the sun is a class G star that on the hotter side of the range, close to the F class. Finally, the V is a roman numeral indicating the sun's luminosity class. Stars fall in different groups on the Hertzsprung-Russell diagram, which plots temperature against luminosity. These groups correspond roughly to the stars' size and lifecycle, ranging from supergiants (I) down to white dwarfs (VII). The sun's V luminosity class indicates that it is a main sequence star.

15 THE CALIBRATION PROCESS

Calibration exposures

Now that we have a solid understanding of image files and color, we can explore the first step in image processing: calibration. This process synthesizes multiple exposures into one image while correcting for defects in the sensor and optical system. By stacking many hours of total exposure time into one image, we can nearly achieve the signal-to-noise ratio of a single long exposure.

To adjust for the characteristics and noise of the sensor and optical system, we take calibration frames (usually flats, darks, and bias frames). As with light frames, the more calibration frames, the better. Each type of exposure is described below. We'll cover the details of how to take them later.

Light frames are the images of a deep-sky target. In calibration, these exposures are corrected using the calibration images.

Flat frames are exposures of an evenly illuminated field taken with the optical system in exactly the same physical configuration as was used for the light frames. They are designed to characterize the light falloff from vignetting, the effects of any dust shadows on the sensor, as well as the variation in photosite sensitivity. All optical systems have some degree of vignetting, with the fastest optics generally having more than slower optics. Commonly used illumination fields include a twilight sky, a t-shirt over the objective, a carefully lit indoor wall, a white LCD panel, or a specially designed luminescent panel. The exposures should be in the middle of the histogram where the sensor's response is linear. DSLR users can use the camera's aperture-priority setting for flats (usually "Av" on the dial). In calibration, the light frames are essentially divided by an average flat frame to yield an evenly lit image.

Dark frames are exposures taken with no light reaching the sensor, and they usually match the light frames' duration and temperature. The goal is to record the thermal signal and account for hot and cold pixels. By matching the exposure time and temperature of the light frames, the average thermal signal can be subtracted. Because the thermal signal is linearly proportional to the exposure duration and temperature, some calibration software can adjust dark frames that do not match the light frames.

Hot pixels are the most visible manifestations of thermal signal. In the enlargement of a four-minute dark frame shown in Figure 78, the black background pixels have a 16-bit ADU level of less than 250, while the visible hot pixels are as high as 23,000.

Figure 78. Enlarged area of a dark frame

Bias frames (also called **offset frames**) are like dark frames shot with the shortest possible exposure (or preferably with the shutter closed). They capture what the camera records without any signal, either light or heat, registering. This allows us to establish the camera's bias signal and read noise. The average bias is subtracted from the other frames to establish a true zero point.

Flat darks are a dark frame for the flat frames. They are shot at the same exposure and gain as the flats, but with the lens cap on or shutter closed. They are not typically necessary unless your camera is uncooled and your flats are very long (>0.5 second). Subtracting the bias frame is nearly equivalent for most cameras. If the flat frames were shot at a different gain setting than the lights, then separate flat darks with matching gain are recommended.

The table summarizes the characteristics of each type of calibration frames. Where settings are usually the same as the light frames, this is highlighted in blue.

Each set of calibration frames is synthesized into its own master calibration frame. By averaging many exposures into one master calibration frame for each set, we increase the certainty (reduce the noise) in our estimate of each pixel's value. Given how easy they are to take, there's no reason not to have plenty of each. The number of calibration frames is not dependent on the number of light frames. Signal-to-noise ratio follows the same square root function

	Flats	Darks	Bias	Flat darks
Exposure time	Until histogram peak is near middle, but keep short	Same as lights	As short as possible	Same as flats
Gain	Use flat darks if different from lights. Lower gain can yield lower noise.	Same as lights	Same as lights	Same as flats
Temperature		Generally the same as lights, unless software can scale darks	Generally not important to closely match lights	Same as flats
Target	Evenly illuminated field	No light	No light	No light
Shutter/ Optical train	Open, in same configuration as lights	Closed	Closed	Closed

for darks and flats as it does for lights (e.g., taking four times as many exposures will cut the noise in half). The same rule applies for all exposure types: the more the better, but with diminishing returns.

The master bias is usually then subtracted from the master dark so that only the thermal signal is left. Some calibration software can then scale the final master dark to approximate the signal at other temperatures. Either the master bias or the master flat-dark is subtracted from the master flat. The process is outlined in Figure 79.

Once the master calibration frames have been created, each light frame can be calibrated. The master bias and master dark are subtracted from each light frame, and each light frame is divided by the master flat. This is outlined in Figure 80.

Once each light frame is calibrated, they need to be aligned with one another to account for any shifts that took place between exposures. The alignment process corrects for this by specifying how each of the subexposures line up against a reference subexposure (the sharpest one is usually chosen for this). These adjustments are made to an accuracy of a fraction of a pixel. Three transformation values are

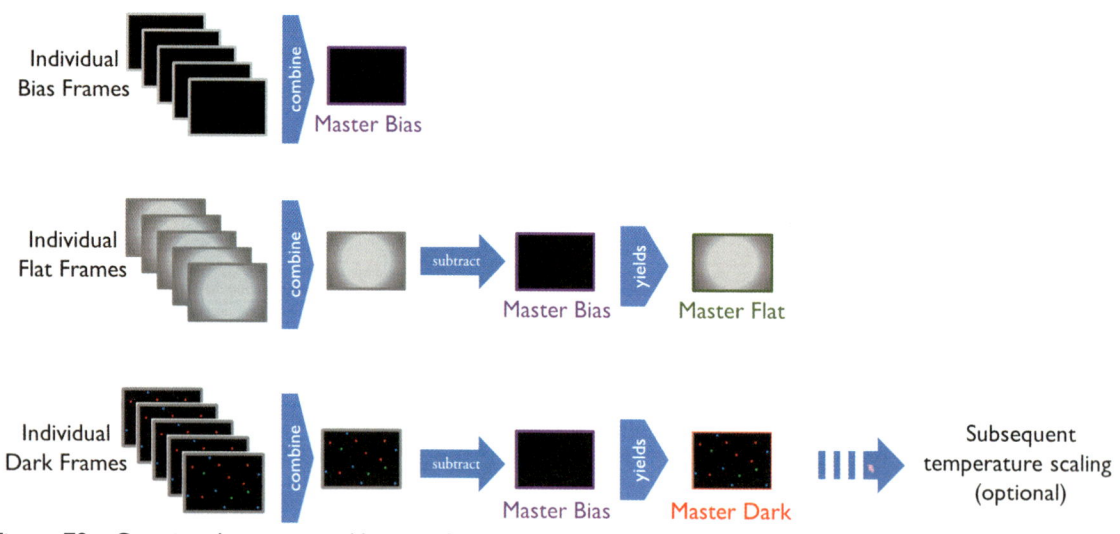

Figure 79. Creating the master calibration frames

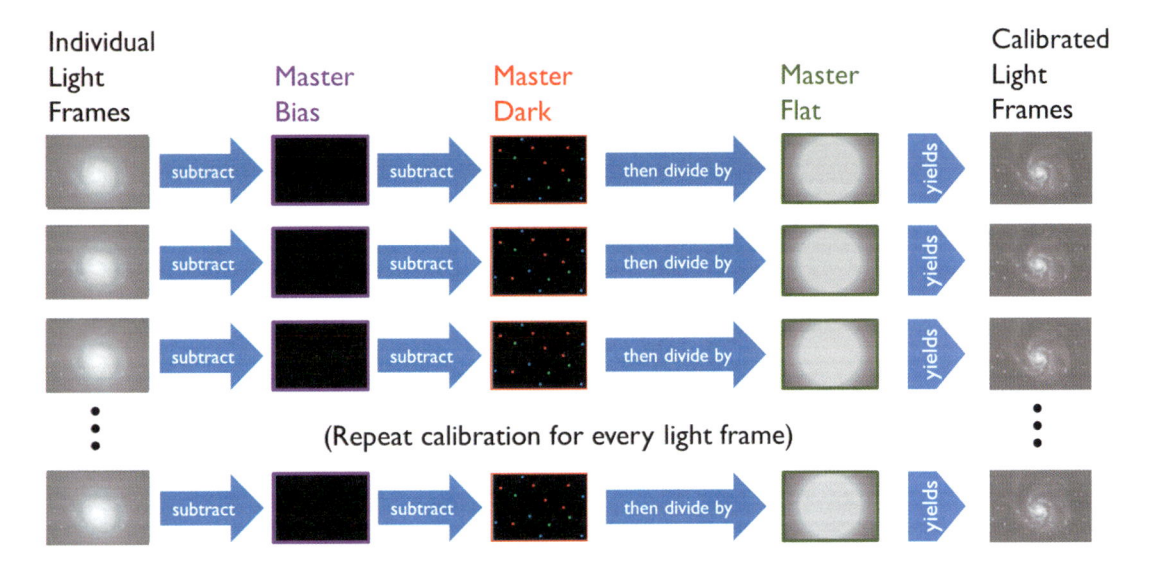

Figure 80. Calibrating and stacking light frames

calculated for each subexposure relative to the reference image: an angular rotation around the center of the image, a shift in the x-axis, and a shift in the y-axis. When applied, these transformations align all of the subexposures. Figure 81 illustrates a simplified example.

Fortunately, all of this heavy mathematical lifting is done for us behind the scenes by the calibration software. Once all of the calibration and alignment is done, light frames are finally ready to be "stacked," as in Figure 82. This will average the calibrated light frames to improve the SNR, but as we'll see, there are more sophisticated options than a straightforward average.

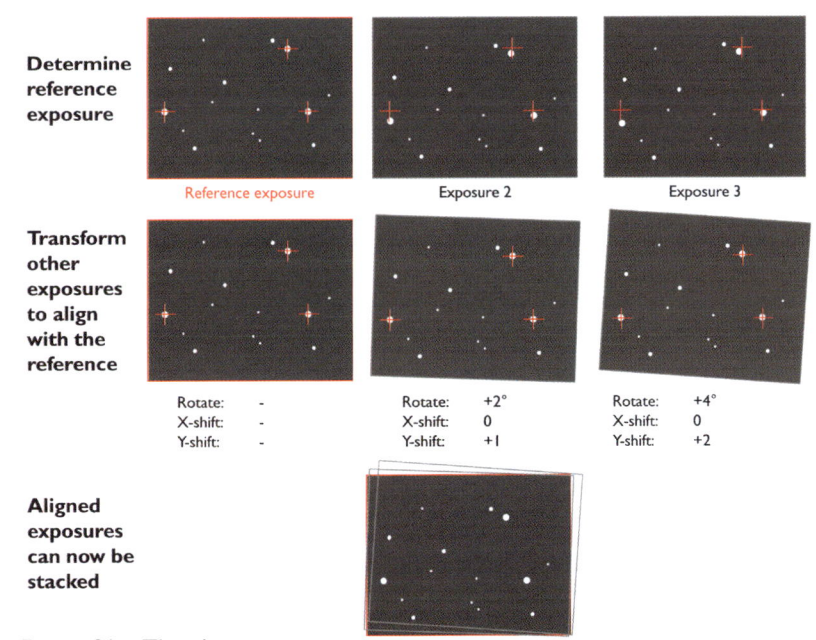

Figure 81. The alignment process

Calibrated Light Frames

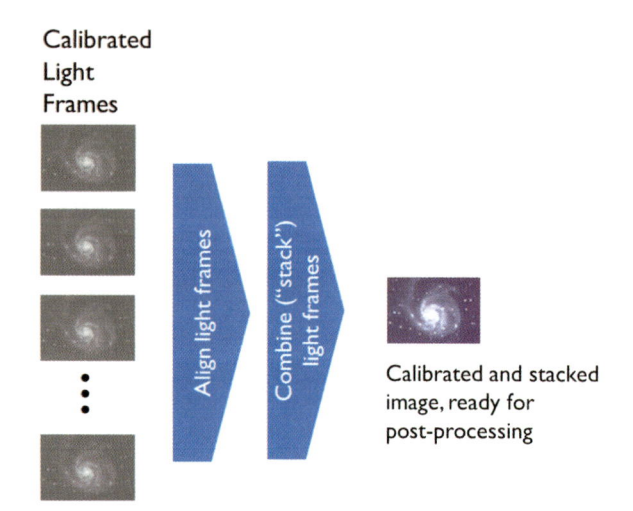

Align light frames → Combine ("stack") light frames → Calibrated and stacked image, ready for post-processing

Figure 82. After calibration and alignment, exposures can be combined

> For uncooled cameras like DSLRs or even cameras with unregulated cooling, it can be difficult to match the temperatures of dark and light frames. Most DSLRs have a sensor that records the camera's internal temperature in the images' Exif metadata. There are several programs available to help you manage your dark library that can automatically select darks with matching temperatures.

Stacking parameters

The choices for stacking method can sound daunting. With so many choices, what method is best? This was a more difficult question a decade ago when users had to consider computational time in their decision. Simpler methods like mean or median stacking were faster, and they produced good results. But with memory capacities measured in gigabytes and multi-core processors, sophisticated statistical methods can be applied to even the largest files in a matter of minutes.

Simple mean stacking. Taking the average or mean value for each pixel from every exposure is the most fundamental method of stacking. It takes full advantage of all the data available, yielding the lowest possible noise, but a common problem is that satellites, meteors, cosmic rays, and airplanes can all show up in the final image, even if they were only in one subexposure. Even averaged with 99 exposures where a particular background pixel is nearly black, the one exposure where it is saturated by an airplane's lights will still influence the mean. The final image of a straightforward mean stack will have remnants of every satellite, meteor, and airplane trail. While these can be processed out by hand later, it is easier to use more sophisticated variations of mean stacking that analyze each pixel's values and exclude the outlier values without throwing out the entire exposure.

Simple median stacking. The benefit of taking the median value in a data set is that it reduces the influence of outlier values compared to taking the mean. This helps exclude any satellite trails or other unwanted intrusions. There are few reasons to choose a median method of stacking, however.

Bear in mind that the goal of stacking is to get the most accurate possible estimate of the true brightness level of the sky at each pixel. Using an averaging method allows us to make a fractional estimate that falls between discrete integer values—the result of averaging can have a greater bit depth than the input values. Data are lost with a simple median method, and gradients in the image will not be as smooth.

As a thought experiment, consider a 3-bit camera. This camera can resolve only eight levels of brightness (0–7). The true brightness of any pixel certainly falls somewhere between two of these discrete values, but we know that we don't have the bit depth to resolve these differences in a single exposure. By averaging enough exposures at our computer (which can manipulate nearly any bit depth we ask it to), we can make a precise estimate of the true brightness, and the accuracy will improve as we add more data. Perhaps with 20 exposures, our average is 5.82 on the 0–7 3-bit scale, and with 50 exposures, we refine our estimate to 5.816. To express this accurately as an integer value takes 16 bits. With a median method, no matter how many exposures we take, our estimate is either 5 or 6, with the possibility of sometimes getting a median of 5.5 if there is an even number of exposures. Basically, we're limited to a 3-bit (occasionally 4-bit) estimate. Consider how a smooth gradient would look in each of these scenarios. With the median stack, there would be clear stair-stepping in brightness (posterization), no matter how many subexposures are used. With an averaging method, the smoothness of the gradient is only limited by the number of exposures taken and the noise characteristics of the system.

An argument could be made that with the 16-bit data most cameras capture, the differences between median and mean stacking are not as pronounced. While this is partially true, consider that most of the data in an astronomical image lies within a very small range of brightnesses. If you are trying to tease apart sixteen levels of brightness (which is a very common scenario, especially with narrowband data), that is effectively like dealing with four bits. Given the effort it takes to collect sufficient data for a great image, why sacrifice any accuracy to median methods?

Sigma clipping. Sigma clipping, also known as kappa-sigma clipping, excludes all values more than a specified number (kappa) of standard deviations (sigmas) from the mean value for each pixel, then averages the remaining values. If you have plenty of exposures, it usually makes sense to exclude values two or more standard deviations from the mean (kappa = 2). If you only have a few exposures, a setting like that may exclude only the most extreme values. Since airplanes and satellites are typically many standard deviations brighter than the background, this can be a simple solution for small stacks.

A variation on this is median kappa-sigma, which replaces outlier values with the median value. This results in a slightly more robust average, though it pushes the average closer to the median. In practice, there is little difference, but using median kappa-sigma is likely to improve results when you have fewer exposures. Both are excellent all-purpose methods for most images.

One major caveat to any sigma method is that the variation in brightness between images should be small. If the imaging run contains images shot through a range of light pollution levels or moon phases, the images should be normalized—rescaled in brightness to match a reference—to allow the algorithm to accurately distinguish what levels are truly outliers. (This is done in DeepSkyStacker with the somewhat obliquely named "background calibration" setting. The "per channel background calibration" version is generally the one to use.)

Other statistical methods. There are simple algorithms, like min-max, which excludes only the single highest and lowest values. For stacks of only a few exposures this should almost always be avoided. Even for larger stacks, consider something more sophisticated. It may do a good job of excluding airplanes and satellites, but it won't work as well if five of the 25 exposures were affected by a passing haze, and the minimum does not always need to be thrown out.

Software like DeepSkyStacker also provides more advanced methods for stacking images. In particular, auto-adaptive weighted average calculates a mean and standard deviation, as in the kappa-sigma methods. However, instead of excluding or replacing the outliers, it weights values closer to the mean higher and those further out less. The weighted average is then determined. Depending on the parameters used, results may not be substantially different from median kappa-sigma, but it never fully excludes any value from the average.

For high-dynamic range images (where the difference between the brightest and dimmest details of interest is extremely large), there are many adjustments that can be made in post-processing as will be described later, but there are also calibration methods that can help before then. Entropy weighted average is one such method that robustly combines images with different exposure times or great differences between the darkest and lightest regions.

Calibration also includes a lot of cleaning up. Every sensor has hot and cold pixels that need to be corrected. Calibration software can do this automatically by interpolating a value from the surrounding pixels. For most sensors, the default hot and cold pixel correction settings will work well. More complex sensor defects like full-column defects may not be fixed by all calibration software, in which case this will have to be done manually in post-processing.

> DSLR users may be tempted to use in-camera dark subtraction, which is usually a menu option called something like "Long Exposure Noise Reduction." There are two major reasons you should not use this feature for astronomical imaging. First, it only subtracts a single dark frame, which is not nearly sufficient, especially at warm temperatures. Second, since it takes the dark exposure immediately after the light, you are cutting the total exposure time in half. For most situations, that time would be better spent (SNR improved) capturing more photons from the target object. Dark frames can be take on cloudy nights or during the day, so there is no need to waste precious clear sky time on darks.

An example of calibration and stacking with DeepSkyStacker

The following example brings together exposures of M106 taken over two nights. While this example uses DeepSkyStacker, the basic process and principles will be similar with any other calibration software. All of the images were 10 minute 2×2 binned exposures taken with the same telescope (f=1625 mm at f/8) and camera. Red, green, and blue filtered exposures were taken, and here we will process the green exposures. Due to intervening weather, the images were taken on two nights six days apart, May 11 and May 17. Separate flat frames were taken after each night of imaging for each filter. Dark and bias frames were drawn from a library of existing images.

DeepSkyStacker's interface has three main sections, shown in Figure 83. The left pane contains the major calibration and stacking procedures laid out from top to bottom in the order they are applied. The bottom pane lists all of the files and their details. There are tabs along the bottom that allow images to be grouped together. We'll use this feature since we have separate sets of flat frames for each night. The biggest pane on the top-right shows a preview of the selected image file.

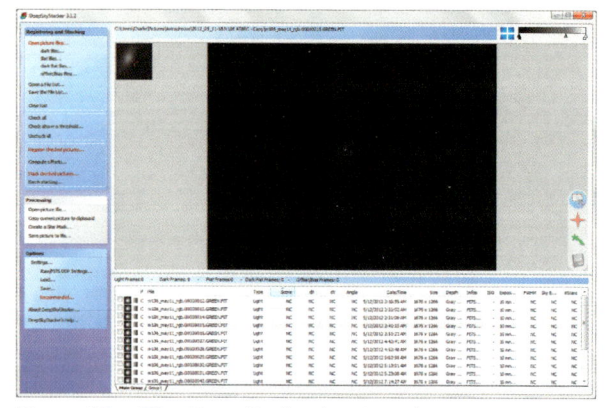

Figure 83. The DeepSkyStacker interface

First, we'll open the light frames from the first night, May 11 (Figure 84). This is the first option in the left pane, "Open picture files." Select the desired files and click "Open" to add them to the file list. In each column of the bottom pane, DeepSkyStacker lists various attributes of each file. Many of these are filled in now, but some are calculated later in the registration phase. Have a look at these to verify the exposure settings are consistent—time, gain, dimensions, etc. The light frames are all unchecked by default. They will be checked later as we decide which ones to stack.

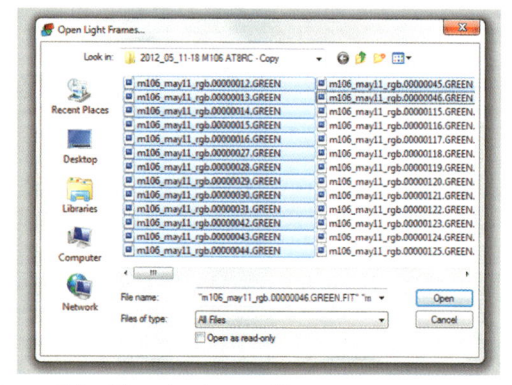

Figure 84. Opening image files

As the preview pane shows, M106 is nearly invisible in these uncalibrated, unstretched images.

To add the second set of images from May 17, we want them to be in a separate group since they have their own flat frames. Click the "Group 1" tab at the bottom of the file list to open the second group. Add the additional images as before using "Open picture files" (Figure 85).

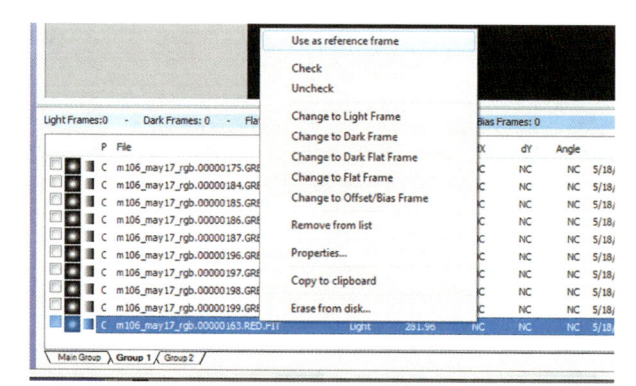

Figure 85. Using groups and setting the reference light

In addition to this second set of images, there is one special file to add. Because we will later combine red, green, and blue channels together into a final color image, we need all of the individual stacks to align with one another. The best image from the red exposures was arbitrarily selected as the reference frame for alignment of all three sets. This image is added to the file list, then it is manually defined as the reference frame. Do this by right clicking on it and selecting "Use as reference frame." If you don't manually define a reference frame, DeepSkyStacker will automatically set the highest scoring image as the reference frame if you do not manually set one.

Now we need to add the calibration files. Under "Open picture files" in the left pane are options to add the calibration files. Since all exposures are the same duration and resolution from the same temperature-controlled camera, we'll use the same dark and bias exposures for both groups. They will only appear in one of the groups. The bias file is an existing master bias created from 100 2×2 binned exposures.

DeepSkyStacker is a little more limited than other software in its ability to group sets of calibration files, but for a situation like this it functions well. Calibration files are checked by default when they are added to the file list. The program will warn you if the calibration files do not match the lights in resolution or another major parameter.

We have two sets of flat exposures, one for each night. Select the appropriate tab at the bottom of the file list and add the flats for that night to each. We won't be using dark flats since the flats were such short exposures.

Click "Check all" in the left pane. We'll consider all of the exposures at this point, then eliminate the ones we don't want later. Now click "Compute offsets." The program will automatically create the necessary master calibration frames, calibrate the light frames, and determine the off-

Processing Images

sets necessary for final alignment with the reference frame (Figure 86).

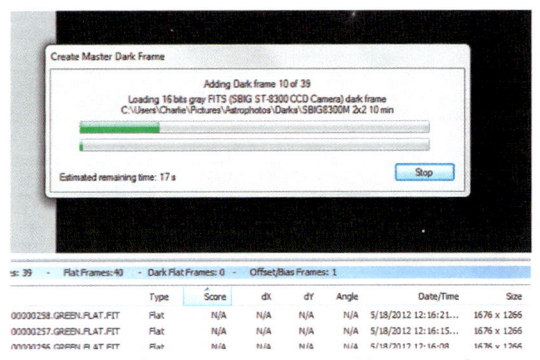

Figure 86. Creating master calibration frames and computing offsets

All of the columns in the file list are now filled in, so we can see how each aligns with the reference, as well as information about the quality of each image. This will help us determine which exposures to keep and which to exclude. Since we are using a reference frame that was taken through a red filter, we need to uncheck it so it is not included in the final stack. It will still be used for alignment.

Uncheck any light frames that are low quality. DeepSkyStacker offers a useful image quality score to help, and you can use "Check above a threshold" to automatically select all images with scores higher than a specified threshold. It's still worthwhile to manually examine the lower scoring images to determine if there are any major defects. Streaked stars from mount errors can still score highly. The score is based primarily on the number of stars found in the image, so you can't compare scores between objects—widefield images with lots of stars will have very high scores, whereas this image at 1625 mm contains only a few discrete stars to begin with. Check the sky background level to exclude any brighter images (= more noise) where perhaps a cloud passed over. Also look at the FWHM (Full Width at Half Maximum) values to evaluate the tightness of stars in each image. These exposures were very good overall, so we'll keep 24 of the 27 green-filtered exposures taken across the two nights (Figure 87).

Figure 87. Excluding low quality light frames

Clicking "Stack checked pictures" brings up the stacking parameters dialog (Figure 88). Notice that there are two steps outlined, one for each group we created. DeepSkyStacker will also provide recommendations for appropriate settings based on the number of files.

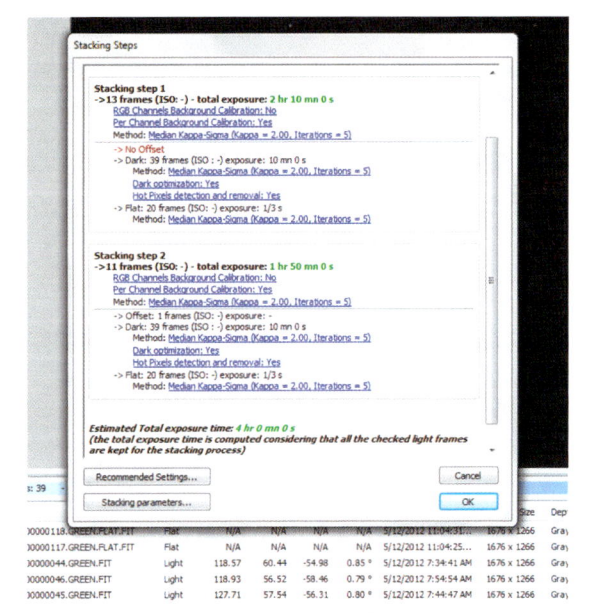

Figure 88. Getting ready to stack

Opening the "Stacking parameters" window via the button on the lower left allows you to adjust any of the stacking options. There are several tabs, each with a group of related parameters. The first tab has options for the final output, including the drizzle method, which is covered in a separate section. Subsequent tabs allow you to set the stacking method for lights and each type of calibration frame, and set options to save any intermediate files. The cosmetic options tab contains the parameters for hot and cold pixel cleaning.

Once you have everything set the way you want it, let it stack. The processing time depends on the number of images, their resolution, and the speed of your computer, but it usually takes only a few minutes. In the final image (Figure 89), M106 is clearly visible. A slight misalignment between frames is visible on the left edge. Once all of the color channels are combined, this can be trimmed.

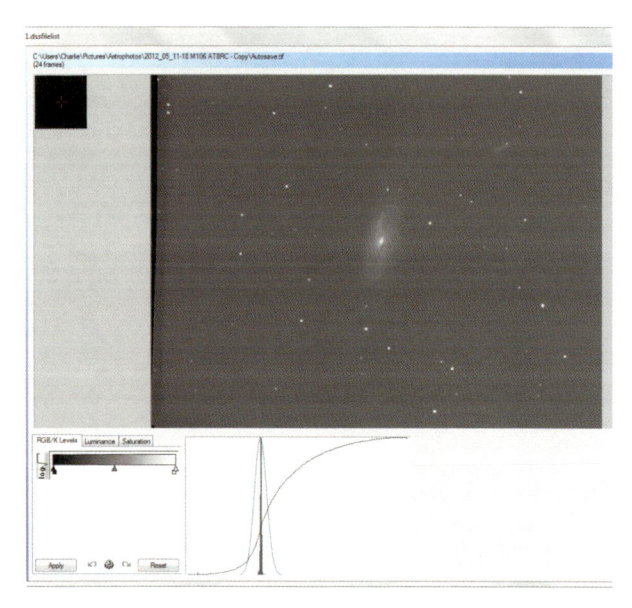

Figure 89. The resulting stacked image of M106

Figure 90. LAX to JFK via the Iris Nebula?

Clicking "Save picture to file" saves the stacked image. The "Save as type" drop-down box contains options for file format and bit depth. DeepSkyStacker provides some basic tools for stretching the image, but this is better done in other applications.

For a composite color image, the stacking process must be repeated for each color channel, taking care to use the same alignment reference frame for each. These could be images taken through red, green, and blue filters or through narrowband filters. If there will be a separate luminance channel, that image is also calibrated and stacked on its own.

Throwing out problem shots
Before stacking, it's worth a quick look through the exposures to weed out any with serious problems. Mount errors, wind gusts, or passing clouds can affect the final result, but other problems will disappear in calibration. Image quality scores catch most of the major errors, but nothing beats personal judgment in finding exposures that could lower the quality of the final image.

It is common to find airplane lights (Figure 90), streaks from satellites and meteors, or tiny gamma ray spots in long exposure images. Because these are large deviations in brightness compared to the other exposures, these images do not need to be discarded as long as you are using one of the stacking methods that excludes outlier values (kappa-sigma, median, min-max, etc). These will exclude such defects while still capturing valuable data for every other pixel in the image.

The most important things to check for before calibration are tracking errors and background brightness. Images with oblong stars will have a detrimental effect on the final image. The tracking error acts as a point spread function over the whole image, not just the stars, so star correction in post-processing will not help the rest of the image. The background brightness can also signal trouble for an exposure. Brighter exposures can be the result of a passing cloud, increasing humidity, a rising moon, or new light pollution from a nearby source. These increases in skyglow add shot noise to the final image, so if a subset of exposures are substantially brighter than the others, it is usually worthwhile to exclude them.

While the general rule is that the more exposures the better, that assumes equal quality for each exposure. If you have 80 equal-length exposures of a nebula, but 40 of them were suboptimal because of light from a crescent moon (or high thermal noise, or haze, or any other imaging bugaboo), and 10 were affected by wind that caused minor tracking issues, you will probably get a better result stacking the best 30 than by stacking all 80. Garbage in, garbage out.

Aligning monochrome images for color combination later
When capturing filtered monochromatic images that will later be synthesized into a color image, each filtered set needs to be separately calibrated, but each of these sets must be aligned with each other as well. One way to accomplish this is to simply calibrate and stack each set separately, then run an alignment on the three resulting image. Another way is what was done in the calibration example above: bring the reference frame from the first set into the subsequent sets, defining it manually as the reference frame for alignment, but excluding its data in those stacks.

The drizzle algorithm

"Drizzle" is an up-sampling algorithm developed by Andrew Fruchter and Richard Hook in the 1990's for use with the Hubble Space Telescope. For small targets or targets with fine detail, stacking the exposures with the drizzle method can yield a resolution greater than that of the camera, though never greater than the atmospheric seeing. The basic idea is that if you have frames that are shifted slightly from each other, they can be aligned to sub-pixel accuracy and a super-sampled image can be interpolated. Thus, in order to use the drizzle algorithm, there must be some shift between exposures, even if a fraction of a pixel. Some image capture software can accomplish this by shifting the mount a tiny amount (dithering) between each exposure. In essence the individual subexposures are precisely aligned over a finer pixel grid, typically with two or three times the resolution (four or nine times the number of pixels). The average intensity of each fine pixel is determined from the overlaid portions of the larger pixels of the original subexposures.

Figure 91 is a schematic example of a 2×2 drizzle on a small area of an image. The original images are aligned on a star, and the calculated alignment point at the star's center is marked by a red dot. A virtual image space is created where the pixels are then reduced in size and laid over a grid with twice the resolution. The brightness value from each of the larger pixels is "drizzled" onto the grid of finer pixels in proportion to the amount of overlap. This is repeated for each subexposure, building up a final image with a higher resolution.

Drizzling can be especially useful for revealing detail when your optical system doesn't quite have enough focal length for a small target, or when you've binned pixels to improve signal-to-noise ratio. Because it up-samples your image to one with many times more pixels, drizzle can require large amounts of memory and computing time. If this is a problem for your computer, you can select only a region of your image to be stacked.

It is important to note that you cannot beat the atmosphere with drizzle. Using this method on an optimally sampled or oversampled image will give you a larger file size, but will not reveal any additional details. Drizzle can only improve resolution if the original image is undersampled relative to the atmospheric seeing, and in that case it can help round star shapes and smooth gradients. Figure 92 shows a small region of stars from an undersampled image where significant detail was recovered with the drizzle algorithm.

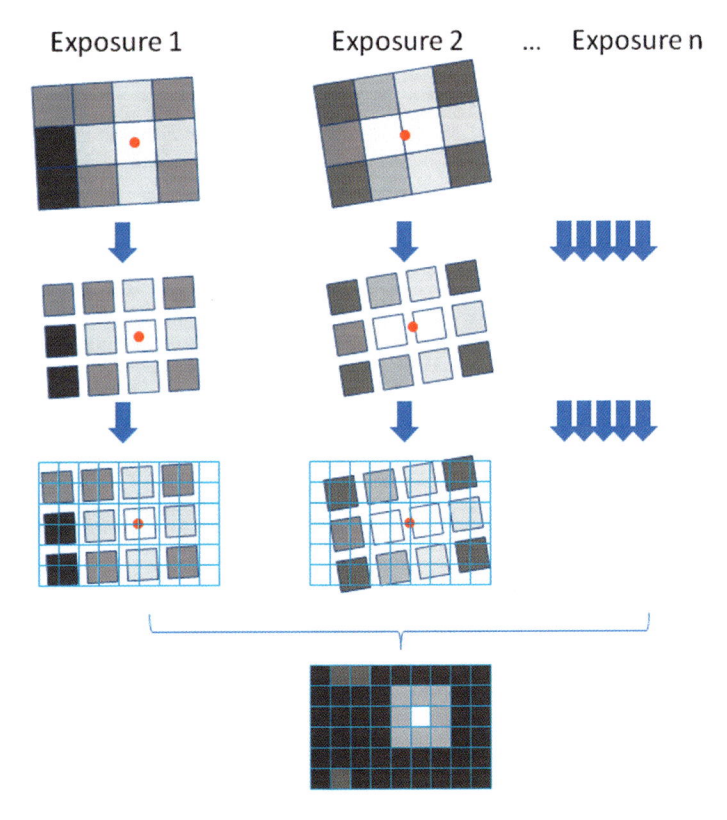

Figure 91. The drizzle stacking method

Drizzle is not blurring or simply rounding off edges—it is going back to the original subexposures and stacking them in a way that captures more spatial information.

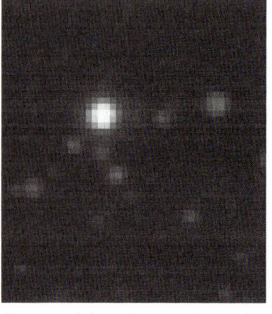

Figure 92. A small region of an undersampled image shown without drizzling (left, 3.38' per pixel) and with 2x drizzling applied in stacking (right, 1.69' per pixel)

Diagnosing defects in calibration output

Dust. When a dust mote moves between exposures, it can cause problems in the final image. Figure 93 shows where a speck on the sensor has moved between shooting the light frames and the flats. The calibrated light frame shows an artificially bright region (left side, above green arrow) where the dust mote sat on the flat frame (right, above green arrow), since that area was brightened in calibration to compensate for the dimness on the flat. The position of the dust mote during the light exposure (blue arrow) remains dimmer than the background since it was uncorrected by the flat.

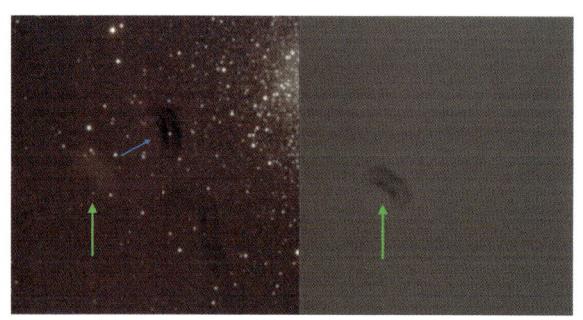

Figure 93. The effect of a dust mote on a calibrated light frame (left) and the corresponding region of the master flat frame (right)

The solution? Check your flats to see if the mote moved while taking them, and exclude those that don't match the lights, then recreate the master flat for a new calibration. If the dust mote moved or disappeared between the time you took the light frames and the flats, you'll have to process the defect out by hand. This is why DSLR imagers should disable automatic sensor cleaning in their cameras. In both cases above, the problem is in an unimportant part of the image, so you could use Photoshop's clone tool, or perhaps burning and dodging to correct it. If it had landed on the main target object, more sophisticated methods would be needed, like creating a synthetic flat from the image, or that part of the image, and dividing it out.

Gradients. Light pollution is probably the most common cause of gradients. As an object moves across the sky, one side of the exposures is closer to some kind of skyglow (the moon, urban light domes, etc), leaving the final calibrated image with a gradient. Wide-field images are particularly susceptible. Gradients can also be caused by flat frame problems. If the sensor's view is not evenly lit in the flat exposures, the calibration process will introduce an artificial gradient in the final image. It is surprisingly easy to make this mistake. This can happen when taking twilight flats with a wide-field telescope where enough of the sky is captured that there is a difference in brightness across the image. It can also happen when a light box is not aligned orthogonally to the scope, or if t-shirt (or other diffuser) flats are taken while a shadow is cast over the diffuser.

Bloated stars. Bloated stars without a brightened background usually mean that focus was lost. This can be from temperature changes or more commonly, focuser slippage. If stars are bloated and the image is hazy with a bright background, it is likely that either thin clouds passed over or dew or frost formed somewhere in the imaging train. The images are unusable in these situations.

Figure 94. Frost is fine, as long it's not on the optics

Hot pixel streaking. If hot pixels are not adjusted before the subframes are aligned and stacked, any motion between frames (especially poor polar alignment) will be revealed as streaks in the final image. The hot pixels have a fixed position on the sensor, so they do not stack on top of each other as the subframes are shifted to align the stars. Debayering of color sensors can result in these streaks taking on a solid red, green, or blue color. If you have these streaks, revisit the hot pixel removal settings in the calibration software. This problem is exacerbated by poor temperature matching of the dark frames with the lights. If a few make it into your final image, they can be fixed manually in post-processing with tools like the Healing Brush in Photoshop or potentially the color selection tool, but it is usually a sign of poor thermal correction by the dark frames.

16 Principals and tools of post-processing

You've collected the photons, calibrated and combined your exposures into one or more stacked images, but the result at this point is usually far from beautiful. Ideally, there would be a cookbook of steps to follow for every image, but this is where art starts to outweigh science. The rest of the book will cover post-processing, and while there are many recipes and variations that can be used, they are unified by a few core concepts and tools.

The learning curve for post-processing astronomical images can be steep. Much like a painter gaining experience with brushes and paint, mastering the tools of the art allows you to implement your aesthetic vision. No one tool is optimal for all tasks, and each object is like a unique scene. Globular clusters require a very different approach to processing than reflection nebulae. The same galaxy may require a different approach to processing in a wide-field image versus a narrower view. There are principles that guide these aesthetic choices, but within the limits of the data, most processing is the result of each imager's unique interpretation of the scene.

Selective adjustments

We can use the tools of image processing globally to the whole image or selectively to parts of it. Selective changes are more powerful, but over-zealous application can lead to an image that is closer to a painting than a photograph. While global changes can be construed as more faithful to reality, consistency is not the same as fidelity. (For example, changing all of the reds to blue in an image is a global change that is hardly faithful.) Some selective changes, like removing color-fringed stars or cleaning up calibration artifacts, simply correct defects in our equipment or process. These actually improve image fidelity. On the other end of the spectrum, painting in some nebulosity that exists nowhere in the image data is just fabrication, much like an airbrushed magazine cover.

Selective changes are applied in four primary ways. The first is by user selection. Most photo editing tools are designed this way—the user has a cursor with which to paint, burn, dodge, blur, sharpen, etc. Or perhaps we select an area of the image with one of the selection wands and apply changes only within that selection. We can also apply changes to specific levels of brightness (e.g. with the Photoshop curves tool) or to specific ranges of color. This separation of color from luminosity is one of the most important principles in astronomical imaging. The final way to apply selective changes is by scale. At a minimum, this can mean separating stars from the rest of the image so the larger structures can be processed on their own. More advanced techniques can separate structures by spatial frequency. This may be the most powerful method for many image adjustments, but it is more difficult to accomplish with standard photo editing software. Some specialized scientific imaging packages like PixInsight and IRIS feature spatial tools like wavelets that allow these manipulations.

This book covers mostly Adobe Photoshop. Version CS3 is used for the processing illustrations and examples, but the tools are nearly the same from CS3 on. In subsequent versions the interface has changed slightly and some new tools were added, but everything shown is possible in the newer versions. While expensive, Photoshop is the de facto image processing standard for other kinds of photography, and it currently appears to be the most popular choice for astronomical image processing as well. The principles described here apply to other software packages, though each will have a different set of functionality used to implement them.

Because each program has its strengths, most imagers use more than one application to process their images. Even the newest version of Photoshop is not complete. Its 32-bit processing is limited, and it lacks spatial frequency tools like wavelets. Photoshop's native sharpening and noise reduction tools are not as good as commercially available plug-ins, which can add to the cost. Combining Photoshop with a dedicated astronomical package (see the Equipment Check for examples) covers the full range of processing tools.

Understanding the histogram

A histogram is a graph showing the relative frequency that specific ranges of values occur. In photography we are almost always referring to a histogram of brightness levels in an image, and it is perhaps the most important visualization tool in image processing. Histograms are fairly intuitive to understand: along the horizontal axis are all the possible pixel brightness values from completely black shadows to completely white highlights, and the number of pixels at each determines the height of the bar on the y-axis. The

Processing Images

histogram tells us about an image's dynamic range and how it is distributed. Much of the work we do on an astronomical image is to stretch the calibrated image's dynamic range so that it fills the full potential range available, and the histogram is our guide in this process.

You can see the histogram in Photoshop via the dedicated histogram panel or as part of the levels or curves tool. It is such an important tool in digital photography that it is displayed on most digital cameras as a way to judge exposure. Note that the scale is not always linear. Photoshop and most imaging software will show a linear histogram,

---

Equipment Check — Software choices for post-processing

There are many software options for both calibration and processing, and any summary of these programs is inherently subjective. The only way to know for sure if an application is right for you is to try it. Below is a sampling of what's currently available.

DeepSkyStacker. Luc Coiffier's DeepSkyStacker freeware has become one of the most popular programs for stacking and calibrating images. DeepSkyStacker does a great job making the calibration process as straightforward as it can be while still providing enough options to allow some flexibility for advanced users. It handles raw files from a multitude of cameras, and it provides useful information about all of the subframes, including a good implementation of a quality scoring system to weed out problem shots. DSS is designed for calibration and stacking, however, not post-processing.

IRIS. Christian Buil's IRIS is also freeware, but provides more manual flexibility in the calibration process, as well as some sophisticated image processing tools. There are few limits to what you can do with IRIS, but it demands a more intimate knowledge of the calibration and stacking process. IRIS provides enormous customization and control through a command-line interface or custom scripts. The program is not frequently updated, as it is a mature program.

Registax. Registax was originally developed for stacking planetary images, and most of its functionality is still aimed at that discipline, but its wavelets functionality can be used for deep-sky images. Its ability to align images can also be used to create mosaics. Registax is freeware.

CCDStack. CCDStack is commercial software that offers one of the most comprehensive sets of calibration and stacking tools available. Every step is tweakable, and it has calibration tools and options found nowhere else.

Nebulosity. Craig Stark's Nebulosity offers excellent camera control features, plus stacking and calibration tools, as well as basic post-processing tools. It can interface with his PHD autoguiding freeware to dither between exposures.

PixInsight. PixInsight is a commercial application that takes a unique "object-oriented" approach to image calibration and processing. For those who want to customize everything and are willing to learn the necessary details, this is a quality software package that is dedicated to scientific imaging, with some very useful features not available elsewhere. This idiosyncratic approach to processing is not for everyone—PixInsight definitely stands away from the rest of the pack with its interface and tools. Then again, a deep mastery of the program may lead to results that are unobtainable from other software.

MaximDL. The full MaximDL suite provides the most complete set of tools for astronomical imaging in one package. Maxim has an impressive array of calibration tools that can automate much of the process, as well as observatory control, image planning tools, and post-processing. A lower-priced version is also available with only the image processing tools included.

Photoshop. Use a recent version of the software, CS3 or later, as these have features very useful to astronomical images that earlier versions lack, like better 16-bit support for tools and filters. As of the time of print, "light" versions such as Photoshop Essentials, do not include sufficient 16-bit tools for astronomical image stretching.

GIMP. The GNU Image Manipulation Program (GIMP or more weirdly, "The GIMP") is a free, open-source raster image editing program designed to compete with Adobe's Photoshop. Two crucial features missing in the 2.x versions are support for 16-bit images and adjustment layers. These are expected to be available in version 3.0. Once 16-bit or higher images are supported, GIMP could be a contender for astronomical image manipulation, especially considering that it's free.

while the back-of-camera histograms are usually non-linear, showing the result after a logarithmic stretch (known as gamma correction) has been applied.

In Photoshop the brightness levels on the histogram go from 0–255 (eight bits), even if the image itself has greater bit depth. Using tools with this 8-bit histogram in no way reduces the bit depth of the image; the histogram simply shows only eight bits of granularity.

Figure 95 shows an image of NGC 281 (the "Pac-Man" Nebula) straight from calibration with its histogram, and then the same image after it has been stretched. The initial histogram shows that all of the image data is squeezed into a narrow range. After processing, that small spike on the histogram is stretched to fill the dynamic range, resulting in a much more pleasing image that reveals the nebula. This histogram looks more like what we would expect from a normal daylight image.

Let's take a closer look at that processed image of NGC 281 to examine how the histogram correlates with the image.

We could divide the histogram into any arbitrary number of ranges, but Figure 96 shows the image split into four segments that illustrate the major components of most astronomical images. Each part is shown on a light green background to provide contrast.

The first segment consists of the left half of the main "peak," which for this image is levels 0–30 on an 8-bit scale. Because the sky background usually takes up most of the image, the first peak on the histogram of most deep-sky images contains the sky background. Any variation in signal down at these low levels is primarily due to noise, so processing here consists of either compressing these levels into a smaller dynamic range or using filters to reduce noise in these areas. The right side of the main peak, levels 31–75 in this example, contains the faintest details of the nebula that rise just above the background at this stage in the processing. The range of brightness from 76–181, the "toe" of the histogram, contains the core of the nebula, the dim stars, and the halos around the brighter stars. Finally, the

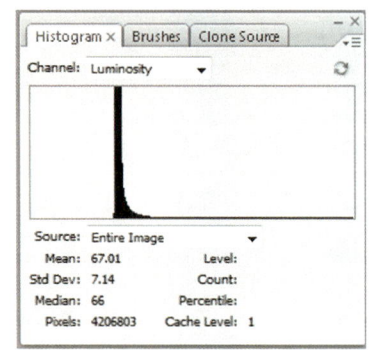

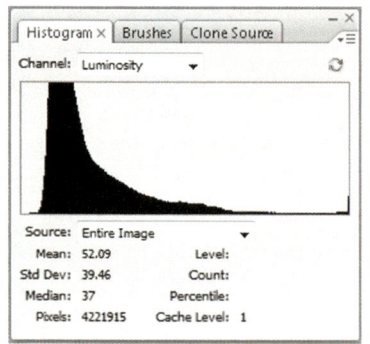

Figure 95. A calibrated but unstretched image of NGC 281 (left) and the same image after stretching (right)

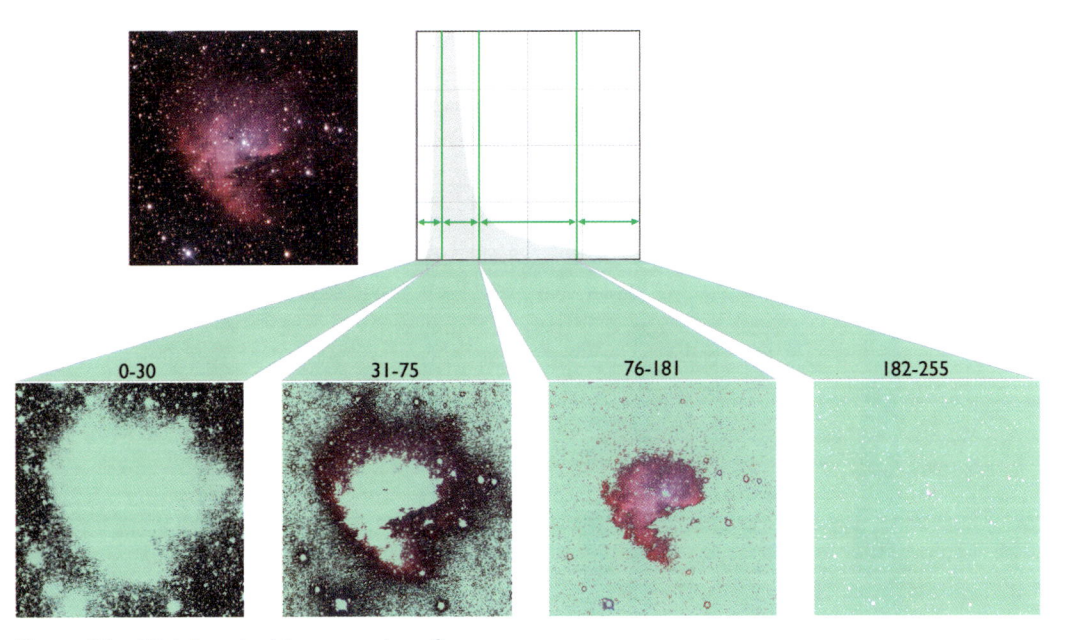

Figure 96. Dividing the histogram into four ranges

fourth and brightest segment from 182–255 contains only the brightest stars.

Given the large range of the fourth segment, there is probably room to further expand the dynamic range of the second and third ranges, which contain image data we are interested in, at the expense of the fourth, which doesn't. This would help reveal more of the faint nebulosity as long as the noise can be controlled. When stretching an image, always understand what part of the region is being expanded and what is being compressed with each adjustment. With thoughtful application of the curves tool, we can reallocate the dynamic range to the parts of the image that are most interesting.

The curves tool

The curves tool allows us to selectively compress and expand each level in the dynamic range. This is done by adjusting the shape of a curve that defines how each level of brightness in the existing image will be mapped to a new level in the resulting image. There are two attributes to consider at each point on a curve: the slope of the curve and its vertical distance from the reference line.

The reference line defines the curve (a line, really) that generates no change to the existing image. It runs from the black point to the white point. When the black and white points are at the corners, that is a 45° angle, which is what Photoshop displays. If you move the black or white point, you have to imagine a line connecting the two as the new reference "line of no change."

Wherever the curve is at an angle steeper than the reference line, those levels will be expanded; that is, the dynamic range here will be mapped across a wider range of values than in the existing image (see Figure 97).

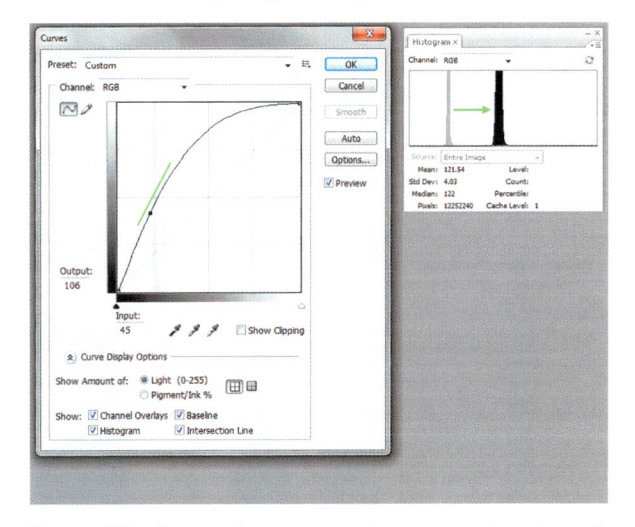

Figure 97. Areas where a curve is steeper than the reference line (in green) results in an expansion of that dynamic range, as seen in the wider spike on the histogram

Where the slope is shallower than the reference line (see Figure 98), these levels will be compressed into a narrower dynamic range. If the curve is horizontal, all of those pixels will be mapped to the same brightness value. Note that downward sloping angles (less than 0°) should never exist at any point on any curve, as they invert levels, swapping light for dark, which aside from making no sense as a processing goal, creates bizarre outline and halo artifacts in an image.

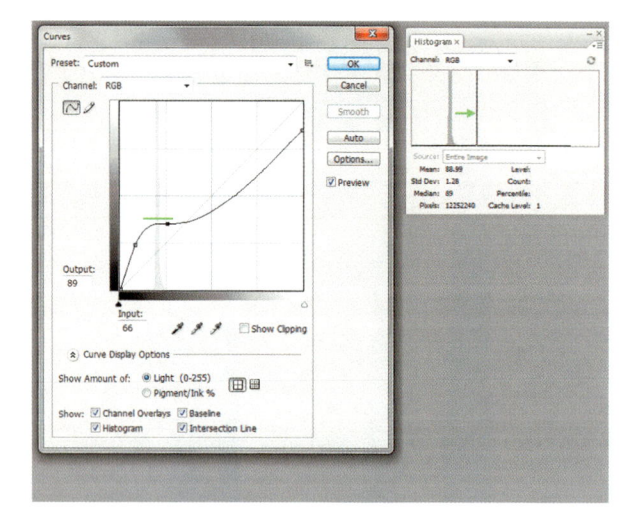

Figure 98. Areas where a curve is shallower than the reference line result in a compression of that dynamic range, seen in the narrower spike on the histogram

Wherever the curve is above the reference line, as in Figure 99, these levels will be mapped to a brighter level, and where the curve is below the line, they will be dimmer in the resulting image.

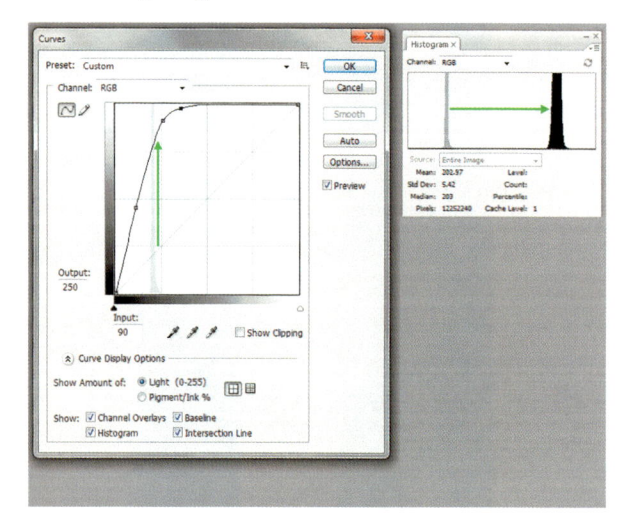

Figure 99. The distance of the curve from the reference line determines how much brighter or dimmer those levels will be in the resulting image. The curve shown leads to a large brightening of the main peak values, shifting them rightward in the resulting histogram, as well as a widening of the peak due to the steep angle. Note that the flat top to the curve has compressed all values past 50% brightness into a single level—this is clipping.

After calibration and stacking, we usually have a narrow spike of image data that we want to stretch across a wider range of values, but we don't want to over-brighten the stars. This is why the first curves used to stretch an image should start out steep and then gently begin to flatten out (but never get horizontal) after the hump that contains most of our image data. The result is to compress most of the stars into a narrower range of brightness values at the top. This makes room to expand our target's dimmer data from the shadows into the midtones.

When stretching an image, be careful not to clip any useful image data. Clipping occurs when you compress a range of brightness levels into one. Once these levels are clipped, you can never recover any distinctions between them—they are all the same value in the new image. Clipping usually happens when setting the black point (the zero brightness value) or the white point (the maximum brightness), which correspond to the left and right ends of the curve respectively.

Figure 100 shows a curve that will clip every value above 77 (in the 8-bit range Photoshop uses for the curves tool), mapping them all to 255, losing all distinctions within that range. While the left part of the curve has revealed the galaxy's arms, the galaxy's core and the brighter stars are all solid white with no details. In a similar manner, starting a curve along the bottom anywhere right of zero will clip shadows.

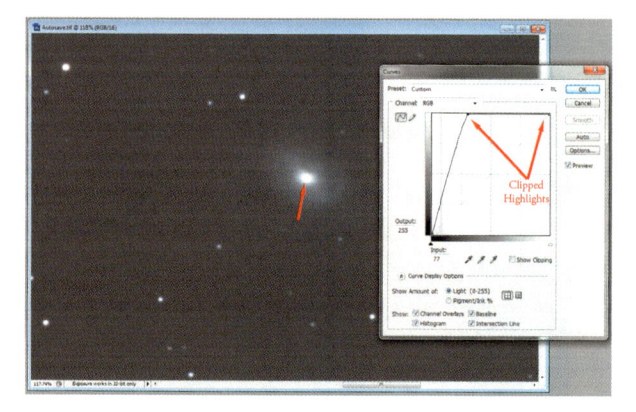

Figure 100. Clipped highlights

While we must avoid clipping any meaningful data, it is important that we eliminate any wasted dynamic range where there are no pixels with those brightness values. These occur at the extreme ends of our image's histogram. By clipping these regions with no data, we can map the brightness values we do have across the entire dynamic range. Ideally, the brightest parts of our image, usually star cores, are at the far right edge of the dynamic range, while the emptiest black space is at the left edge. Since the

Processing Images

data in the deep shadows are frequently below the 'noise floor'—where the variations in brightness are purely due to noise—the black point is usually set right at left edge of the histogram hump.

After the image has been digitally developed with several iterations of basic curves until the astronomical object is visible, we can start to add more complex curves. One of the most useful curves is the S-shaped curve (Figure 101), which will increase contrast. Carefully placing the central inflection point allows you to emphasize specific details in the image, as we'll see later. The idea of an S-curve makes sense when you consider that an S-curve is holding shadows down and brightening highlights, essentially pushing midtones out toward the extremes. (The curve shown is illustrative—it would not be a wise choice for this image, given its histogram.)

Curves should never be arbitrary; they should be based on the content of the image. Always consider the correlation between your image and the regions of the curve. Fortunately, Photoshop provides features that show you how the two relate. When you have the curves dialog open, Photoshop automatically changes your image cursor to the eyedropper. As you mouse over your image, you can see where each area falls on the curve, giving you a good sense of where the shadows and highlights are, as well as any special regions you want to emphasize. If you hold down on Ctrl while you click an area of your image, it will place an anchor on your curve at that level.

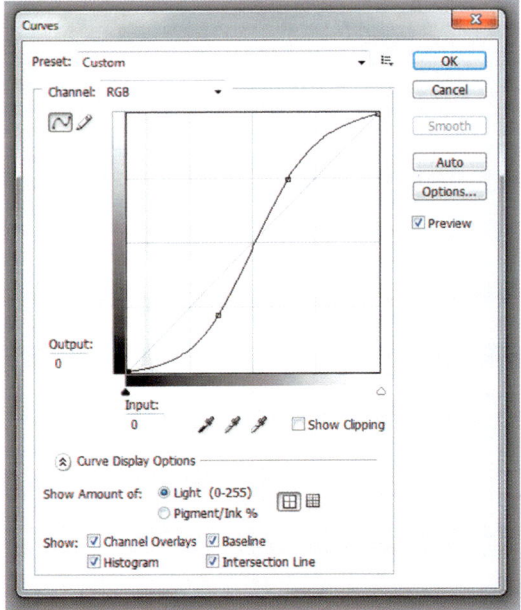

Figure 101. An S-curve creates contrast

Always apply curves iteratively: stretching a little, checking the result, then applying another stretch. There are two reasons for this. First, as we stretch, details in the image are revealed that may need to be accounted for in the shape of subsequent curves. A common example is the slightly dimmer space between the arms of a spiral galaxy. Too much compression between those levels and the levels in the arms can lead to a flat-looking galaxy.

The second reason is that Photoshop has limited resolution in terms of the levels we can put our anchor points. There are only 256 steps within which to pick our points, and one step in the original image may subsequently be mapped to hundreds in the final image. The manipulations we apply in the later stages would amount to fractional step changes in the earlier stages.

> For consistency, PC keyboard shortcuts are listed. If you are using a Mac, Cmd is the equivalent of Ctrl, and Option is the equivalent of Alt.

Levels

The levels tool is a simpler tool than curves, allowing you to set the black point, white point, and a single gray point in the middle. No other shaping of the curve is accommodated. Before 2007 proper stretching in Photoshop required alternating applications of curves and levels because only the levels tool showed the image histogram, which is needed to set the black point without clipping. Since Photoshop CS3 however, the curves tool has incorporated a view of the histogram, and this useful improvement is also part of many other software packages like PixInsight. With this feature there is rarely reason to use levels, since you can set end points at appropriate locations on the histogram via curves. If your processing software's curves tool does not display the histogram, you will have to use levels to reset the black point (and occasionally the white point) after each application of curves.

The midtone slider (also known as the gamma slider) on the levels tool can allow a slightly more sophisticated stretch, but it lacks the control that you have with the curves tool, so it may not be needed if you have curves at your disposal. If you are using Photoshop Elements or another package that doesn't have a curves tool, the midtone slider can be an approximation of basic curves manipulation with one anchor point in the middle. The levels tool is analogous to a curve with only three anchors that you can only move horizontally. This is illustrated by Figure 102, where the curves and levels transformations shown represent approximately equal manipulations.

123

The magic of layers

Layers are one of the most powerful and flexible image processing tools. They are like virtual transparent sheets that allow you to build up an image from multiple sources or, conversely, separate an image into components. The opacity of a layer or any part of a layer can be adjusted between 0% and 100%. The interactions between layers, known as blending modes, can even be defined as something more complex than transparent sheets stacked on top of each other. For instance a layer can be set to draw only the hues from the layers layer below it, using the luminosity from the top layer. We'll review all of these blending modes shortly.

Layer opacity. The simplest way to adjust the interactions between layers is via the opacity slider. After making changes, sometimes we want to "dial back" the effect a little. Lowering the opacity allows the lower layers to show through, mitigating the changes on the top layer.

There is another slider below the opacity slider, called the fill slider. The difference between the two is subtle, but it almost never applies to astronomical images. The fill slider affects transparency exactly the same way as the opacity slider, except it does not affect layer effects. These effects, like drop shadows or beveling, are typically applied to vector and text layers, which are not used in astronomical image processing. Thus for the images considered here, the opacity and fill settings accomplish the same thing.

Layer masks. By masking parts of a layer with varying degrees of opacity, tools like filters that can normally only be applied to the whole image can be masked to reveal their effects only in a selected area. Layers can also let you separate an image into components for independent processing. For many images, the stars stand out as an obvious first choice for isolation to their own layer. They are generally far brighter than the rest of the image, so stretches applied to the whole image can leave them with clipped highlights and a lack of color saturation.

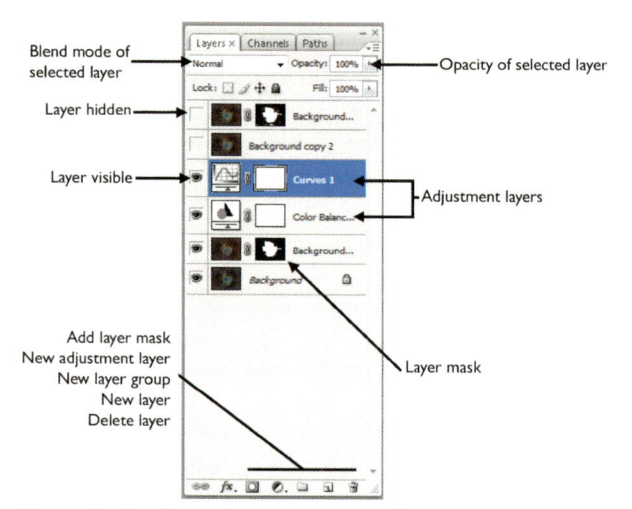

Figure 103. Photoshop's layer panel

Adjustment layers. Layers don't even need to contain image data—a special type known as an adjustment layer contains only information about how to apply a tool or filter to the layers below it without directly altering those layers. These allow you to arrange each processing step as a series of non-destructive layers that can be adjusted individually without permanently altering the original image. Photo-

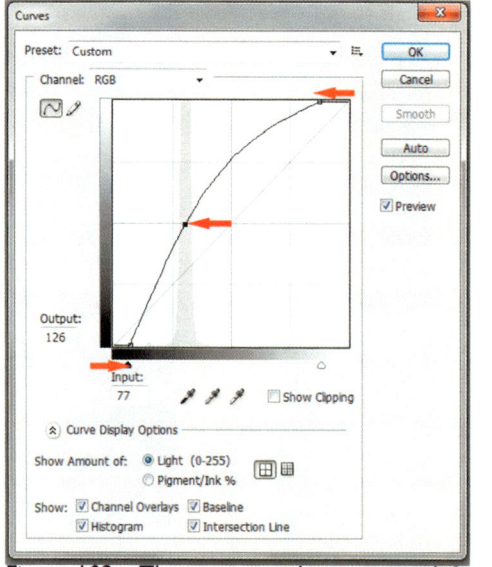

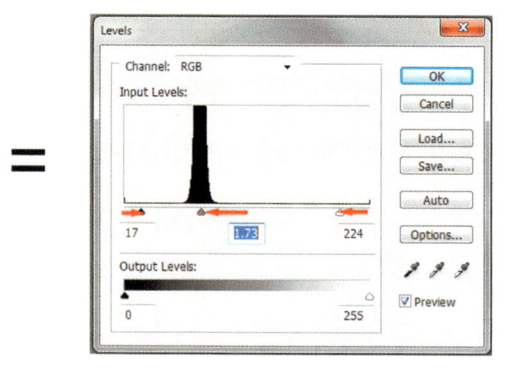

Figure 102. The curves tool can accomplish everything the levels tool can and more

shop 'reads' layers from the bottom up. By default, the changes made by an adjustment layer affect the resulting image at that point in the stack of layers, but an adjustment layer can be set to only apply its effects to the layer directly below via the "clipping mask" option.

> To keep your .psd file sizes small, turn off the "Maximize PSD and PSB File Compatibility" option. It is located under Edit | Preferences | File Handling. With this option on, adjustment layers can dramatically increase the file size.

Layer groups. Layers can even be grouped, allowing you to keep things organized. Simply select the layers you wish to group, then click the 'Create a new group' icon on the bottom of the layers palette. (It's the one that looks like a file folder.) Or you can click the icon to create an empty group, then drag and drop layers into it.

Using layers in your processing is a good habit to start if you haven't already, and we'll use both kinds of layers extensively in the processing examples that follow. When an image is ready for publication, all of the layers can be flattened into a final image, but always keep the layered version in case you want to make changes without starting from scratch.

While not absolutely necessary, mastering the use of layers can allow processing techniques that would otherwise be impossible. The ability to separate an image and even the processing actions applied to it into non-destructive layers is also one of the great strengths of traditional image processing applications like Photoshop that make them hard to abandon.

Layer blending modes

Blending modes offer a more complex way to define the interactions between layers. Photoshop now offers over 20 different blending modes, but only a few are useful for astronomical imaging, and a few of them are hard to imagine using for any image. Blending modes define a mathematical relationship between two adjacent layers based on the values of the corresponding pixels in each. For convenience in describing the modes, the layer whose mode is being set is referred to as the "top" layer here, even if there are layers above it.

In Photoshop's layer math, the values of each channel for each pixel are remapped internally to a decimal value between zero and one. For 8-bit data, zero remains zero, and 255 maps to 1.0. For each corresponding pixel, the values are compared, and a mathematical transformation is applied. Some of the mathematical relationships are simple.

For instance, the Multiply mode simply multiplies the luminance values for each pixel. But the consequences of using normalized numbers can be counter-intuitive. Consider two layers that both have a middle gray value. Applying the Multiply mode would result in the blend layer showing $0.5 \times 0.5 = 0.25$, a darker gray. In fact this mode always results in a darker result, except where both layers are at maximum brightness (1.0), in which case the result is the same level.

Normal and Dissolve. In Photoshop the blend mode drop-down box loosely groups the modes into six groups based on the way they work. Normal is the default mode, and it works like you would expect: any part of a layer that is not transparent covers up the layer below. Dissolve combines the top layer with what's below it by semi-randomly selecting pixels from each.

The Darken Group. The next group, from Darken to Darker Color, consists of modes that all result in a darker image. Darken compares each channel (usually R, G, and B, but it also works with other image modes) of each pixel and takes the darker of the two. Darker Color does the same, but it compares the total brightness of each pixel, instead of working at a channel level. Multiply is commonly used in daylight photography to restore detail to faded or overexposed areas. As we've show, multiplying by pure white will keep the underlying image exactly the same. Anything else will only darken it. Where both layers are dark, the result is dramatically darker, whereas when both layers are light, they tend to maintain their brightness better. Color Burn and Linear Burn produce even more darkening, with greater or less color saturation.

The Lighten Group. The group beginning with Lighten is a mode-for-mode mirror image of the Darken group. Each mode generates higher luminance in the opposite way that

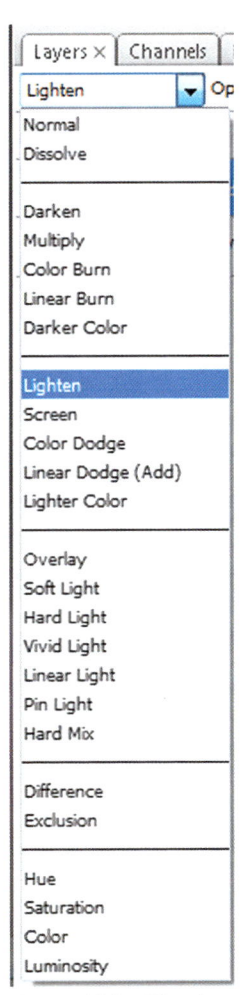

Figure 104. Blend modes (Photoshop CS3)

its complement in the previous group leads to lower luminance. For instance, Screen is the opposite of Multiply. The values of each color channel in each pixel for both layers are inverted then multiplied, then the result is inverted. It sounds complicated, but the effect is simple: a pixel in the resulting image is at least as bright as its counterpart in either of the source layers. Where both layers are light, the result approaches white, and where both layers are dark, the darkening is not as pronounced. The effect is to simulate multiple projectors shining on a screen.

The Contrast Group. The fourth group is the most interesting for our purposes. The modes from Overlay to Hard Mix increase contrast in some way, especially when both layers contain the same or similar image data. They do this the same way an S-curve works: by lowering the shadow values and increasing the highlights. Each mode evaluates whether a given pixel is above or below a middle gray of 0.5. Then, if the value is less than 0.5, one of the modes from the Darken group is applied. If the value is greater, the complementary mode from the Lighten group is applied. Unfortunately, the middle point of the contrast is fixed at 50%, but we will later use this to our advantage by combining these modes with the high pass filter, which keys its output on 50% grey.

Overlay and Soft Light are both combinations of the Multiply and Screen modes (applied to the dark and light values respectively), with subtle variations on how much is applied and which layer is the reference layer. Similarly, Hard Light and Linear Light combine the Linear Burn and Linear Dodge modes. Pin Light and Hard Mix both create extreme results.

The Inversion Group. Difference and Exclusion form the next group, and both invert the image. Mathematically, both subtract the values of each channel in each pixel from each other, then take the absolute value, since negative color values are meaningless. Using Difference, similar colors appear very dark, and different colors and shades appear brighter. With Exclusion, similar colors appear as a middle gray instead. Photoshop CS5 added Subtract and Divide modes, which perform similar mathematical transformations.

The Color Group. Finally, the Hue, Saturation, Color, and Luminosity modes all take the named component of the top layer, but bring in the other attributes from the bottom layer. The Color mode takes the hue and saturation from the top layer, preserving the luminosity from the layers below. Don't confuse this mode with the Hue mode, which takes only the hue from the top layer, using the bottom layers for both saturation and luminosity. The Luminosity mode applies the brightness levels from the top layer, but preserves the hue and saturation from the layers below. This is one of the most useful blending modes, as it can be used to separate and manipulate luminosity independently, even when it isn't in a separate channel.

It is not important to know every blend mode. The result of applying one is rarely intuitive, so it takes some experimentation to understand them, and many are not useful for our purposes. The most commonly used modes in this text are Normal, Overlay, Hard Light, Saturation, Color, and Luminance. With these six, most of the processing tasks used on astronomical images can be accomplished.

Layer masks

Instead of setting the opacity of a whole layer, layer masks allow you to selectively define the opacity of any region on a layer. It's a simple concept that brings tremendous flexibility. Only the luminance of the mask is used by Photoshop, with white defined as opaque, black as transparent, and shades of gray or any other color as the range of semi-transparency between.

A layer mask can be added via the menu or with the "add layer mask" icon. Clicking the icon creates a mask with the default setting to hide the layers below (i.e., the mask is filled solid white). The menu equivalent is Layers > Add Layer Mask > Hide All. To set the mask to reveal the layers below (filled solid black), hold Alt while clicking the icon or choose the Reveal All option on the menu.

To save a step, you can first make a selection on the image, and then click the mask icon. This will reveal only the selection, hiding the rest of image. Holding Alt while clicking the icon will of course do the opposite, hiding the selection and revealing the rest. The menu has equivalent Reveal Selection and Hide Selection options. Rarely will we want hard edges on our selections, since those will lead to harsh transition boundaries, so the selection is usually feathered or a blur is applied to the layer mask to smooth the transition.

The layer mask can be drawn on directly or image data can be pasted onto it. A thumbnail of each mask is shown on the layers panel. To view the layer mask directly, hold down on Alt while clicking the mask. To invert a layer mask, click it and press Ctrl-I.

Creating a composite layer

One potential issue with editing your image via adjustment layers is that some manipulations, like filters, can only be applied directly to one layer, not to the result of multiple layers or to an adjustment layer. One solution is to save a copy of the image, flatten the layers, and then start sub-

Processing Images

sequent work from there. Another approach is to create a new top layer that is the result of the existing layers. While it doesn't have a menu option, Photoshop has a function to do exactly this. It's called Stamp Visible, and it's only available via the keyboard shortcut Shift-Ctrl-Alt-E. It creates a new layer containing what you see on the screen as the result of all existing layers. This new layer is a snapshot in time—if you change any of the underlying layers, it won't change. It is effectively the equivalent to saving a copy of the image, flattening it, then pasting the result as a new layer back into the original. Because of this, composite layers can substantially increase file size and slow down performance for some computers.

Selections and feathering

Photoshop provides a range of tools with which to select a part of an image. These are part of the standard set for any image editing program. The standard lasso tool allows you to hand-draw your selection. There are also shape selection tools, including a rectangle, ellipse, and even a polygon tool. Holding the Shift key while dragging the corner of these tools (or any other similar tool) will lock in a 1:1 aspect ratio: the rectangle becomes a square and the ellipse becomes a circle. Like any tool, the precise attributes can be set through the options bar that usually runs horizontally under the menus. (If you don't see this bar, turn it on with the menus via Window > Options.)

There are dozens of ways to define a selection. The magic wand, quick selection, and magnetic lasso tools try to select areas similar to what is clicked. The Color Range tool (Select > Color Range) allows you to select a specific color or range of colors. As we'll see this is very useful for adjusting astronomical images, especially for adjusting star col-

ors, fixing gradients, and tweaking false-color narrowband images.

Since defining a selection with hard edges usually reveals the selection's boundaries when it is altered or pasted somewhere else, we nearly always feather the edges of a selection. Feathering a selection gradually fades the transparency around the selection's border. This is prevents an abrupt edge from appearing in the image when changes are made to a selection. Figure 106 shows the same selection (from an image of the nebula IC 4177) without feathering, feathered with a 10 pixel radius, and feathered with a 50 pixel radius. The red bar helps illustrate the transparency of the feathered areas.

If you feather a selection with a radius of 10 pixels, you may be surprised (quite reasonably) to find that the transition appears to extend 25 pixels outward and inward from your selection edge. Why does this happen? The short answer is that you can always expect the feathered zone to extend about 2.5 to 3 times the radius you selected in either direction from the selection's edge, (the "marching ants" line).

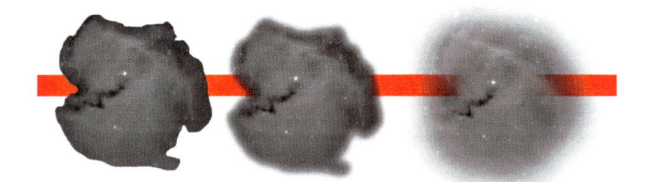

Figure 106. A selection with no feathering, feathered 10 pixels, and feathered 50 pixels.

The more detailed answer is that the transition is not linear; it is computed using a cumulative normal distribution, as

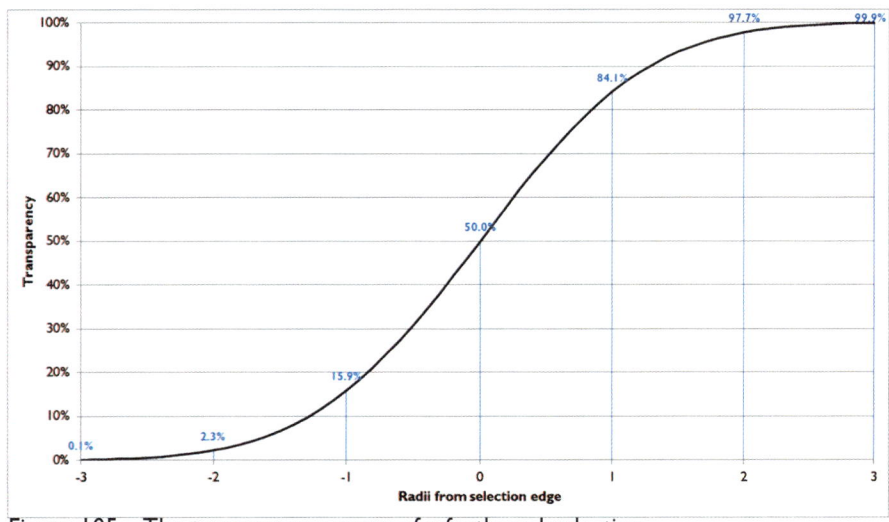

Figure 105. The transparency range of a feathered selection

127

shown in Figure 105. Photoshop sets the edge of the selection as the midpoint of the transition, at 50% transparency. The radius you set is actually the one standard deviation argument for the normal curve formula that defines the transition. This means that the actual transition zone extends around 2.5 to 3 radii on either side of the selection edge. It extends until the computed transparency value rounds to 100% on the outside boundary and zero on the inside boundary. For a radius of one pixel, the feathered zone would extend three pixels on either side, while a 10-pixel feathering extends 25 on either side.

Figure 105 shows the transparency of pixels based on their distance from the selection's edge. The bulk of the feathering happens within one radius of the selection edge. One radius outward, the transparency is 15.9%, and one radius inward, it's 84.1%. Two radii out, the transparency is only 2.3%, and two radii in, it's 97.7%. By three radii, the numbers are 0.1% and 99.9%. By this point, the value rounds to zero for even the largest radius values.

A post-processing workflow

This section describes a basic workflow for astronomical image processing. While everyone will develop his or her own preferred approach to processing, this workflow is a good starting point if you don't have one, and it provides the structure for rest of the text.

Two principles guide the decisions at each point in the workflow: brightness and scale. Brightness is a good indicator of the signal-to-noise ratio. The brightest areas in an image have the highest SNR, so they can be processed most aggressively. This includes stretching, increasing color saturation, and sharpening. Noise is most visible in the dimmest areas since the noise is a larger proportion of the signal. Here we must take steps to reduce the visibility of this noise. For the sky background in particular, we can be more aggressive with our approach, using desaturation and filtering to create a cleaner canvas for the rest of the image. Finally, we must try to visually separate details by creating contrast selectively. The faintest wisps of nebulosity need to be lifted above the sky background, and the subtle variations between midtones in a galaxy's arms need to be exaggerated.

We consider scale because structures at different scales also need different kinds of processing. The more we can separate our processing of small-scale variations versus larger ones, the better we can highlight or suppress features in those spatial realms appropriately. For instance, separating the stars from the rest of an image allows us to apply non-linear stretches tailored to reveal the larger, dimmer features. Applied to the whole image, these curves would otherwise over-compress the stars. Sharpening can then be applied to highlight galactic dust lanes or variations in nebulosity without creating halo artifacts around stars. Photoshop gives us a very limited set of tools for dealing with scale, but software like PixInsight or other packages with wavelet-based tools provide far greater flexibility in this respect.

The workflow that follows is a starting point for most deep-sky images. The exact order and tools used will vary for each image, but nearly all images will go through three main stages: image development, detail enhancement, and final polishing.

Image development is about bringing the dynamic range of the image to a point where the deep-sky object(s) reveal all of their detail and stand out from the background. Most of the processing here is done to the whole image. This stage usually includes:

- Initial stretching (which may include saturation boosts along the way)
- Background adjustments: gradient removal and color neutralization
- Combining color channels to create a color image if the images were taken with a monochrome sensor
- More stretching, with smaller steps focused on specific details

Detail enhancement adjustments are mostly selective, focused on bringing out specific details. This usually entails:

- Creating a separate luminance channel
- Selective sharpening and local contrast enhancement, usually only to luminance
- Selective color adjustments to create accurate color and saturation enhancements to boost it
- Star size reduction
- Cosmetic repairs – artifact and defect removal

Final polishing involves the last adjustments to clean up an image and get it ready for presentation:

- Fine gradient removal
- Noise reduction
- Cropping and orientation

The following chapters are loosely mapped to this sequence of processing steps, however some actions are performed at multiple stages on an image or the sequence varies.

17 STRETCHING: REALLOCATING THE DYNAMIC RANGE

Non-linear stretching

When we "stretch" an image, our goal is to reallocate its dynamic range to reveal the most detail in our target object. This involves expanding the data in the shadow and midtone levels across a wider range of brightness while compressing the bright levels into a narrower range. At each end of the histogram, we also want to clip any wasted dynamic range where there are no data.

What is the difference between a non-linear stretch and a linear stretch? A linear stretch involves only adjusting the white point and/or the black point. In the Curves tool, it looks like a straight line, as in the left side of Figure 107. A linear stretch clips the extreme ends of the dynamic range, expanding the range of the remaining data in the middle linearly. Mathematically, the relationship between each level of brightness between the two points is scaled by a linear factor.

If any changes are made to the way dynamic range is allocated within these middle values, it is a non-linear stretch. In Photoshop this would include any adjustment of the midtone slider in the Levels tool or any adjustments in the Curves tool where an anchor point is not on one of the outside edges or the 45° reference line.

In order to reveal most deep sky objects, significant compression must occur in the brighter regions to free up enough dynamic range to expand the shadows. Managing these trade-offs is the job of non-linear curves, and their judicious application constitutes the primary aesthetic choice (and a lot of the work) in post-processing.

We'll apply multiple rounds of gentle non-linear curves, building up our changes rather than doing it all at once. We use an iterative approach for several important reasons: First, as we stretch an image, we reveal details or regions that we can take into account when determining the shape of the next curve. As the dynamic range of the object expands, it becomes clear exactly what part of the histogram corresponds to what part of the image. Second, if you are using Photoshop for post-processing, its tools only have eight bits (256 levels) of resolution with which to apply adjustments. The underlying image typically has at least 16 bits of data (65,536 levels), and we aim to ultimately make precise adjustments that require more precision than an 8-bit scale affords on the initial image. A single curve, no matter how complex, is not sufficient to accurately stretch an image, so we proceed through an iterative process of curve applications. Finally, an iterative approach allows us to apply other adjustments, such as saturation enhancements, as we go.

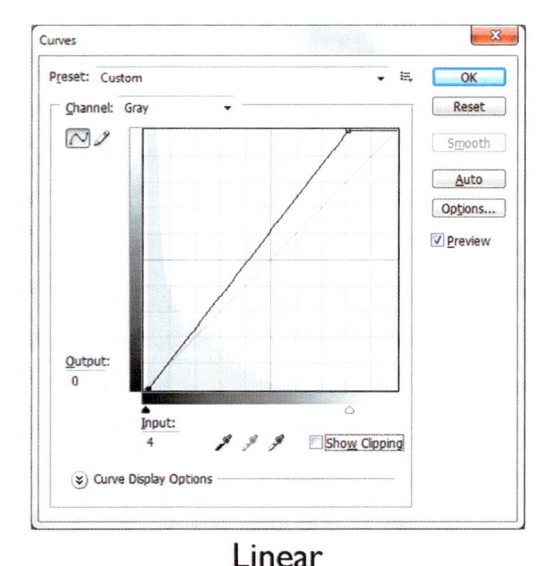

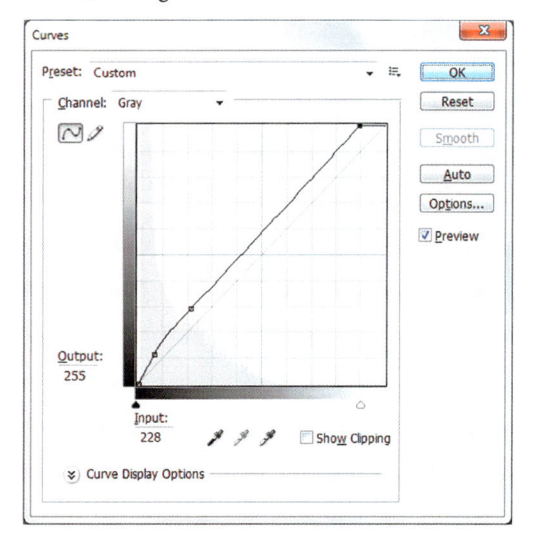

Figure 107. Linear and non-linear curves

Start with balanced colors

If the color channels were not balanced after calibration and stacking, now is the time to do it. One way to do this is to set up some Color Sampler points in the image so you can monitor the color balance of the background as you stretch the image. The Color Sampler tool lies under the standard eyedropper tool in Photoshop. Access it by right-clicking on the eyedropper. This tool allows you to set up to four spots on the image whose color information will be shown on the Info panel (revealed by F8 if you don't have it up). We'll make the assumption that the background, since it contains no objects, should be neutral in color—the R, G, and B values should all be nearly the same. While this assumption does not guarantee accurate deep-sky colors, it is usually a good start.

Set a few Color Sampler points in the empty parts of the background. To reduce the impact of noise, use the 3×3 or 5×5 average setting, and zoom in close to make sure you're not in a region of nebulosity or picking up a star. As you apply curves, keep an eye on the sampled RGB values. If one color channel is out of line with the others, apply a stretch or black point adjustment to that color channel to bring its histogram in line with the others.

Now that we can watch the color balance as we stretch, let's look at an example. We'll use an image of M51, the Whirlpool Galaxy, to demonstrate the three stages of stretching: initial development, object-background separation, and intra-object separation.

Initial development

Use the menu (Adjustments > Curves) or Ctrl-M to bring up the Curves tool. The goal of the initial round of curves is to expand the object's dynamic range—to widen the right half of that spike on the histogram where most of the object data are. At this early point, the main object may not even be visible. Once the spike is wide enough, we can apply curves that selectively affect the image data more than the background.

With all this expansion, something has to give, and it's the brightest pixels, which are mostly star cores. Figure 108 shows that nearly all of the initial image data lie in the left quarter of the dynamic range, with the vast majority confined to a narrow spike. Figure 109 shows the result after the first curve is applied, and a similar curve being applied again. The green area of the curve is steeper than 45°, expanding the left quarter of levels, while the blue part is shallower than 45°, compressing the right half of the histogram. All levels in the image will be brightened except the very darkest and brightest, as the entire curve is above the reference line.

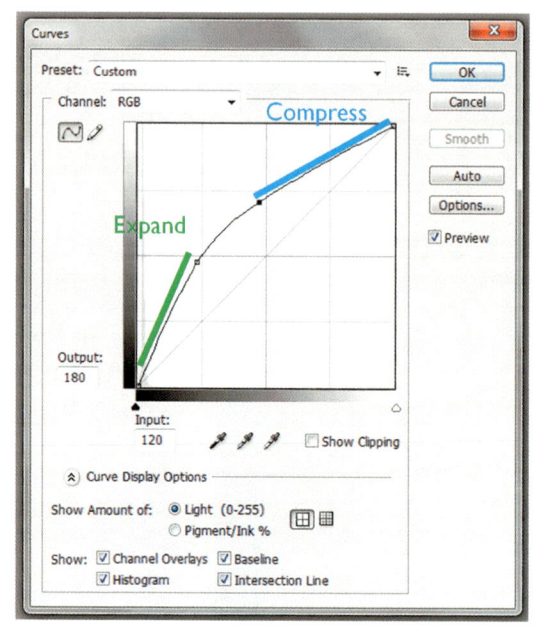

Figure 108. An initial stretch

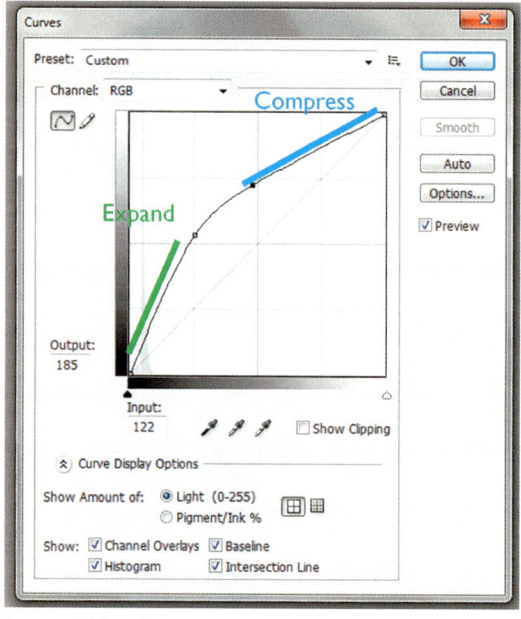

Figure 109. A second stretch (note the change in the histogram)

> There is nothing special about mathematical function curves like ln(), log(), or asinh(). Fixed mathematical curves do not know anything about your specific image: how wide the dynamic range is, where in that range the object's data lies, or what levels you want to emphasize. It is always better to use custom curves with carefully chosen anchor points, applied in multiple iterations. By iteratively using the curves tool to adjust the details in your image, you can effectively building up a very complex final curve shape that is customized to your image's data.

Object-background separation curves

After the image development of the initial curves, the histogram is wide enough to start better targeting our curves. While we want to bring out the faintest wisps of our object, everything dimmer than that is the sky background. We want the background to take up as little dynamic range as possible and remain as dim as possible, so we'll shape our curves appropriately.

After a few iterations of initial curves, the image looks something like Figure 110. Three distinct brightness levels have been highlighted in the image, and corresponding anchor points have been set. Recall that holding the Ctrl key while clicking on a point in the image will set an anchor point on the curve.

A representative spot of the background is shown at the point of the red arrow. As we would expect, it falls in the left half of the main spike on the histogram. Also note that there is a gap of wasted dynamic range between the spike and the left edge. The orange arrow points to the galaxy arm, which we'll eventually want to be a mid-tone in the final image. Note how little separation there is between it and the background. Finally, the purple arrow points to the galaxy's core, which should be a highlight in our final image, though we want to retain some detail here.

Using the background level anchor as a pivot point, we drag the left anchor of the curve along the bottom, clipping most of that wasted dynamic range (Figure 111). This will map the darkest black to the first level where we have data: the left edge of the histogram spike. Since the curve lies below the 45° reference line to the left of this pivot point, we are darkening the background. After the pivot point, which is where the galaxy's data lie, we are lightening everything (the curve is above the reference line). We are also expanding the histogram for most of the range, but compressing it for areas brighter than the galaxy core. This separates the galaxy's arms from the background, making them more like mid-tones in the image. Another anchor point could be used to adjust the inflection between compression and expansion based on your aesthetic preferences.

Intra-object contrast curves

Once the object is starting to take up sufficient dynamic range—once the histogram spike is fat—the goal is to visually separate the brightest and dimmest regions of an object from each other. Creating contrast like this usually involves subtle "micro-curves." This works on a larger scale, for instance separating the arms of a spiral galaxy, and later we'll explore other methods of creating contrast at selective spatial scales.

After another iteration of the previous curve, we have something like Figure 112. To make M51 less flat, we enhance the contrast between the brightest part of the spiral arms and the dimmer region between them. Set anchor points at the level of an arm (orange) and at the level of an intervening dust lane (red).

Figure 113 shows the contrast curve. Holding the dimmer region constant (red arrow), we raise the brightness of the

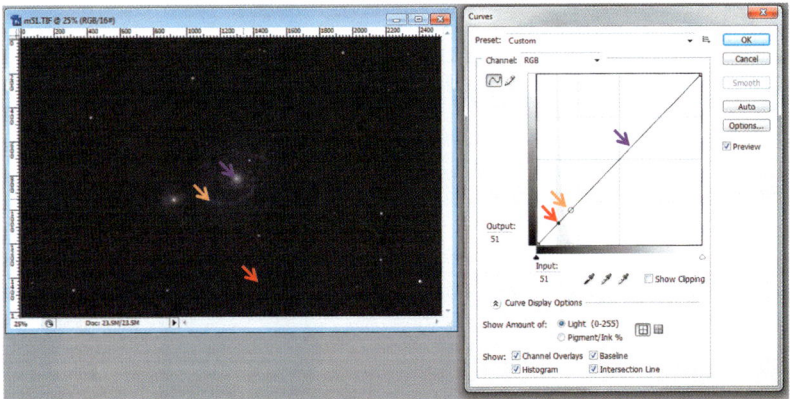

Figure 110. The image after a few initial curves

arms (orange arrow) carefully with the keyboard's arrow, forming a gentle S-shaped curve. Recall that S-curves enhance contrast, but the anchors determine the inflection point for that contrast. We could just as easily enhance the contrast between other areas, like the wispy nebulosity around M51's companion galaxy, NGC 5195, and the background with different points. As always, more anchors can be used to control specific regions. Here, another anchor near the top would help hold down the brightest areas and prevent stars from becoming bloated.

> You can apply curves and most other processing steps either directly to the image or via adjustment layers. Using adjustment layers makes it easy to adjust or undo each step separately later.

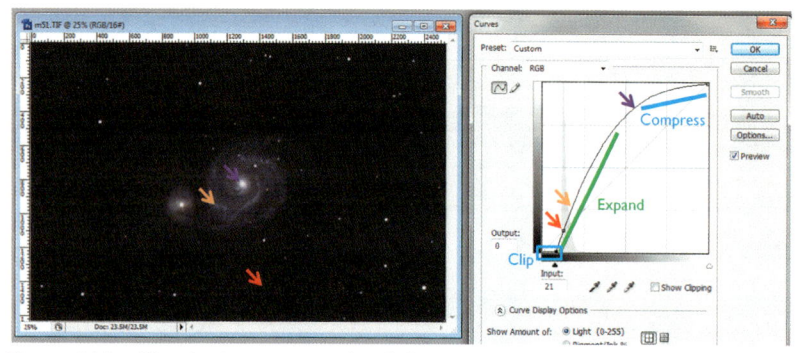

Figure 111. Continuing to stretch while resetting the black point

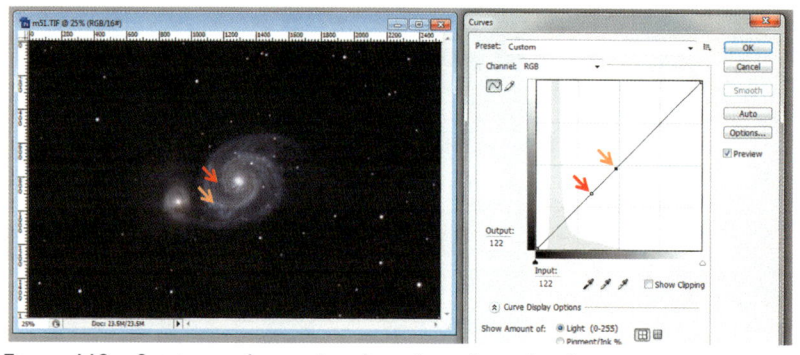

Figure 112. Setting anchor points based on the galaxy's arms

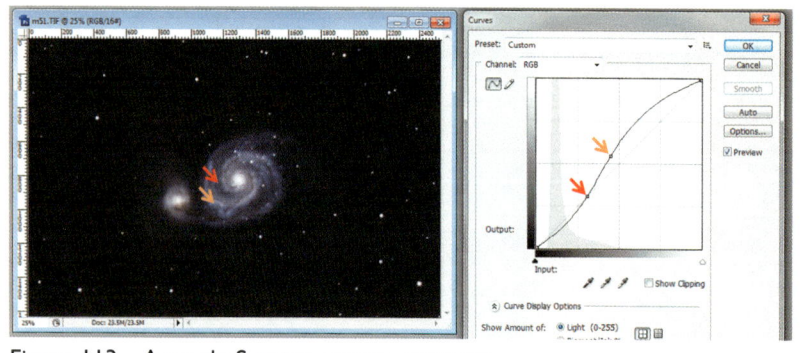

Figure 113. A gentle S-curve creates contrast

132

Processing Images

Controlling highlights and boosting saturation

Galaxy cores or other bright highlights will lose detail if they get too bright or their dynamic range becomes too compressed. As you stretch it's important to pay close attention to these areas. As with most processing methods, there are a couple of approaches to this issue. The fastest is simply to select the bright area, and then use curves to push the brightness back down. The magic wand tool or normal lasso selection tool can be used to make the selection, but be sure to feather it. Do this periodically as you stretch. (Repeated feathering of selection can be tedious to do with the menus, so use the Alt-Ctrl-D keyboard shortcut to bring up the Feather Selection dialog.)

As we'll see later in this section, stretching an image also drains it of color. If you are stretching a color image, it may help to periodically boost the saturation as you stretch. It is fine to do saturation enhancements later in one step, but an occasional boost while stretching may save you from tedious adjustments later. Keep the saturation tweaks small at this point, just enough to maintain color. The simplest method for repeated application like this is to use the Hue/Saturation tool (Image > Adjustments > Hue/Saturation or Ctrl-U), but we'll explore other saturation boosting methods later.

Posterization: the perils of over-stretching

Of course we can't expand the dynamic range of our object forever, but what happens if we try? When stretching a small range of brightness levels across a wider dynamic range, we run the risk of over-stretching to the point where there are gaps between levels—in other words, there are levels of brightness with no pixels at those levels. This is known as posterization.

Imagine an 8-bit image with most of the object of interest's data squeezed into the 10 levels from 46 to 55, with the remainder of the dynamic range (the two hundred levels from 56–255) nearly empty. We are mostly interested in that 46–55 range, so we use curves to expand it across the levels from 46 to 155, 11 times the range it previously occupied. This requires us to compress the less important levels into half their previous range, now from 156 to 255. The problem with stretching that 10-level range across 110 levels is that there were only 10 levels of brightness to begin with. Stretching them can map them to a wider range, but there are now gaps between them. In fact there are now 10 empty levels between each used level in the new image. This points out the main reason we strive to capture as much bit depth as possible, but even 16-bit images can be stretched to the point of posterization.

Because digital images contain only discrete values, they are all inherently posterized to some degree, but our eyes only notice it when it is severe or when our mind expects a subtle gradient. In an image, posterization looks like the opposite of a smooth gradient: there are stair-step jumps in brightness. Posterization also leaves a signature "comb" pattern of vertical spikes separated by spaces on the histogram. Both of these are shown in Figure 114. When the histogram starts to look like this, you've reached the limit of stretching.

The best way to avoid posterization is to start with data with sufficient bit depth and dynamic range to stand up

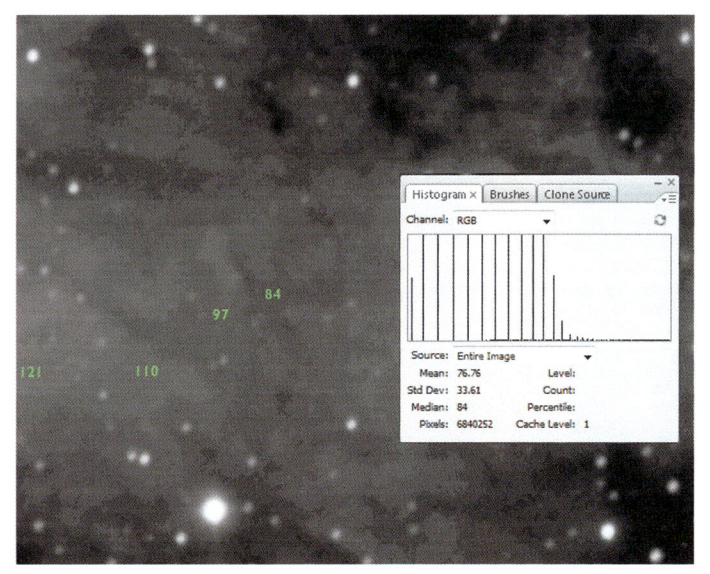

Figure 114. Posterization in a narrowband image (8-bit values for selected areas shown in green text)

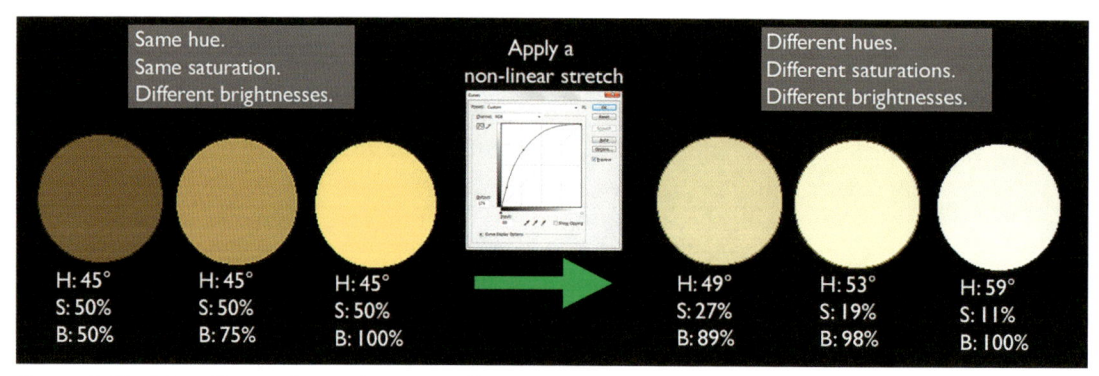

Figure 115. The effects of stretching on color

to the aggressive stretching necessary for astronomical images. While sixteen bits is about the greatest bit depth current digital sensors can capture in a single image, averaging many images together through calibration and stacking can result in an output that require even greater bit depth to accurately describe. But greater bit depth is only worth having if the data you've captured are spread across a reasonable dynamic range. If all of an object's data are squeezed into a narrow spike on the histogram covering only a few discrete levels, those few levels will pull apart from each other when stretched. A histogram that is too narrow usually means that there is too little data—the subexposures were too short, the total duration of exposure time was too short, or the stacking method rounded values somewhere.

The effect of stretching on color

Since the stretching process compresses the dynamic range of the brightest levels, they can lose their color saturation. Stars usually occupy the brightest parts of the histogram, making them particularly vulnerable to becoming washed out. Figure 115 shows an example of stretching's effect on color using three yellow circles that stand in for highly magnified stars. Each has the same hue and saturation, but at a different brightness. Since it's easiest to speak about saturation in the HSB color space, the hue, saturation, and brightness of each circle is shown. The same curves transformation is applied to all three, with the result shown on the right. Note that there is no clipping in the curves transformation, but the highlights are strongly compressed. This approximates the sum of several initial curves applied iteratively.

All three circles get brighter, but the decline in saturation is proportional to the initial brightness. This is because the brightest values underwent the most compression of dynamic range—the curve was flattest in those higher levels. This compression drives the individual color channel values closer together, reducing saturation.

Not only does the saturation change after a stretch, the hue also changes. The original HSB hue value of 45° is different in each of the three resulting colors. What is going on here? Figure 116 shows the same process applied to a fully saturated image of the spectrum. The resulting spectrum loses the gradual transitions between colors.

The color channel values that are high will quickly saturate, so the closer any given color is to a primary or secondary color—red, green, blue, cyan, magenta, or yellow—the sooner it will simply become that hue as the individual channel values are maxed out. (The exact result here, with thicker and thinner color bands, is specific to the hues of the initial image, which was designed to show visual perception of color.)

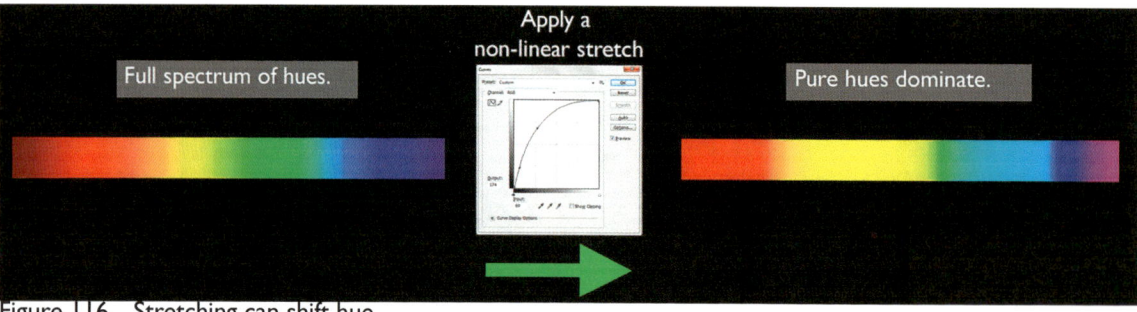

Figure 116. Stretching can shift hue

Digital Development Process (DDP)

The Digital Development Process is a digital process for "developing" images that was designed to mimic some of the desirable qualities of film photography. It was created by Kunihiko Okano in the 1990's, and it does two main things: it applies a non-linear stretch using a hyperbolic curve and it sharpens the image to create edge emphasis. The curve has two parameters to adjust each end of its 'S' shape, but there is always a linear region between these two tails, similar to the response curve of film. The edge-emphasis is also inherent to the way film processing works, and it is digitally mimicked through a process similar to unsharp masking. Some processing software packages have DDP functions that implement a version of the process. It can be a good start toward achieving a pleasing image, especially in terms of color saturation and quick histogram stretching.

Exercises

3.1 What is the color of an image after a curve is applied that consists of a straight horizontal line at the 8-bit value of 200?

3.2 Does using the Levels tool instead of the Curves tool to stretch an image prevent the loss of color fidelity and saturation?

18 BACKGROUND ADJUSTMENTS AND COSMETIC REPAIRS

Gradient removal

Vignetting and dust spots are generally corrected well in calibration, but sometimes stretching reveals a gradient across the image. Most gradients run from one side of an image to the other because they typically occur when one side of the sky in the field of view is brighter than the other. Sometimes there is also a color gradient, where the overall background hue changes across the image without a significant change in brightness. There can even be some remaining vignetting from poor calibration.

There are many ways to remove gradients, and they fall into two groups. The first is to use a dedicated tool. These calculate a two-dimensional background map using polynomial fitting, then subtract or divide this map onto the image. The second approach is to use standard raster editing tools like Photoshop to adjust the background. A gradient is a subtle change across a large spatial scale. Finer scale variations are usually nebulosity that you want to preserve. With either approach, there is a risk of removing faint details from the image if the tools are not applied carefully. With either method, work iteratively to progressively remove any remaining gradient and evaluate the effect on dim features.

Dedicated gradient tools. Both IRIS and PixInsight have powerful and flexible tools for gradient removal. Both allow you to choose the points on the image from which to calculate the background model. They can also show you the resulting model so you can verify that only skyglow is removed. PixInsight's tool is called DynamicBackgroundExtraction, shown in Figure 117.

A commercial plug-in for Photoshop called GradientXTerminator offers a simple user interface without picking points on the image. Created by Russell Croman, this plug-in corrects gradients in a full image or within any selection in Photoshop. Defining a selection allows you to keep the corrections confined to only what it truly the background in your image.

Without a tool specifically designed for gradient removal, there are several techniques for correcting gradients in Photoshop.

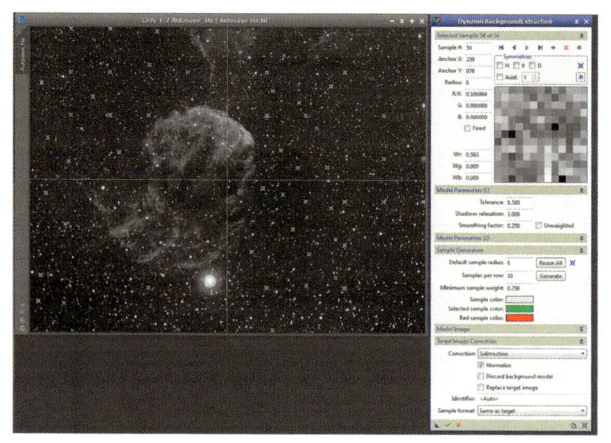

Figure 117. PixInsight's DynamicBackgroundExtraction tool

Correcting color-only gradients. To remove color gradients, you can use the Color Range tool to select the offending pixels directly, and then use curves or the Hue/Saturation tool to adjust their color. Zoom in close on the image in an area with a strong color gradient. Use Select > Color Range to select pixels of the offending color via the single pixel eyedropper tool. Make sure the tool is set for single pixels, not an area. Keep the fuzziness at a low level, then click the additive eyedropper (or hold down on the Shift key while you click) and carefully begin selecting the pixels in the color range you want to adjust. The preview display on the tool will show the pixels with similar color values in the entire image. If it is a true gradient, the selection should be more prominent toward one side of the image. After you've selected ten or so pixels to build a good average, adjust the fuzziness slider to widen or reduce the range of colors you are selecting. If you click on the wrong pixel while accumulating, you can use Edit > Undo or Ctrl-Z to undo the last selection. If you don't catch your mistake before clicking another pixel, you'll need to start over.

Once you have selection, you can use the Curves tool to adjust a single color channel to correct the color gradient by adding the opposing color (if the gradient is cyan, magenta, or yellow) or reducing the offending color (if the gradient is red, green, or blue). Bear in mind that if you push the curve down in one channel, it is usually wise to pull the RGB combined channel up a little to keep the total brightness approximately the same.

Processing Images

Synthetic flats for gradient removal. If we can remove the stars and objects from the image, we can create a synthetic flat that can be subtracted from the original image to flatten the field. Since gradients have a low spatial frequency, we can remove any small details like stars to leave behind an approximation of gradient. This method works well when the objects of interest take up only a small portion of the image and are clearly defined; a galaxies for instance. It does not work as well if the object fills most of the frame or when there are subtle gradients of nebulosity that need to be preserved.

While using synthetic flats is not the preferred method, it can generate cosmetically acceptable results in a pinch. Here, we'll use an image of the M86 region of the Virgo galaxy cluster with a severe gradient (Figure 118).

Figure 118. Part of the Virgo cluster with a gradient

The first step is to duplicate the image to a new layer. In this example we'll work with an image in a single layer, but if yours is composed of multiple layers, you'll need to make a copy of the current resulting "top layer" (what you see on the screen). A composite stamp layer (Shift-Ctrl-Alt-E) works well for this purpose.

Remove the stars with the Dust and Scratches filter (Figure 119), using any of the methods described in Chapter 21. Adjust the sliders until all but the largest stars have disappeared. Here, the galaxies are still visible, but we will remove them in the next step.

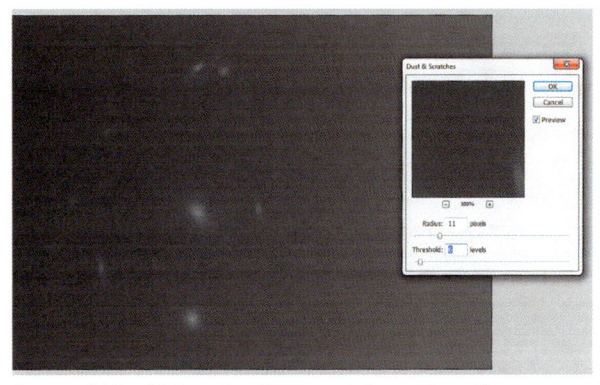

Figure 119. Using the Dust and Scratches filter to remove stars

Removing the galaxies by hand is tedious. A better way is to use the Curves tool to perform a special kind of clipping that will replace any objects brighter than the background with the background level, as shown in Figure 120. Because this gradient is so strong, the background level behind the different galaxies is substantially different, so we'll remove them in a couple of steps. The group of galaxies on the right is on the brighter side of the gradient, so we'll remove those together. Draw a loose selection around them, then open the Curves tool. We want no part of this selection to be brighter than the brightest shade in the background in our selection, so we set an anchor point by holding Ctrl while clicking the brightest part of the background gradient (not part of the object). Now we'll draw a curve that keeps anything dimmer than this as it is, but brings any pixels that are brighter down to that level. This curve follows the 45° reference line up to the anchor point, then proceeds horizontally from there. It is important to keep the 45° part of the line from bowing outward. It should be a perfectly straight line so we don't change the background gradient at all.

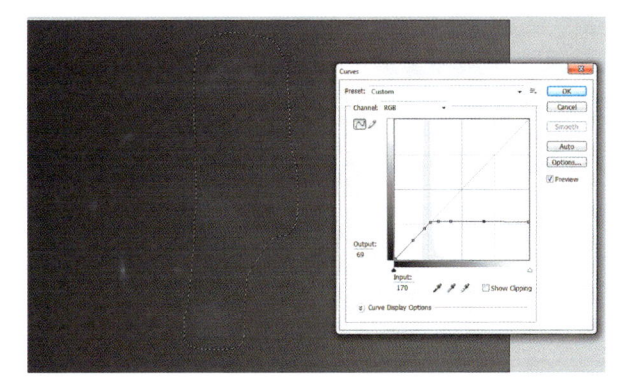

Figure 120. Using Curves to remove galaxies

The galaxies on the left can now be removed in the same manner via a new selection(not shown).

Now that we have our rough gradient map, any final manual adjustments can be made, and then we blur it with the Gaussian Blur filter (Figure 121).

Figure 121. Blur the remaining image

To subtract this gradient map from the image, we simply set the layer blending mode to Difference. This cancels the gradient, but it leaves us with an unnaturally black background that seems noisy in some areas. Blurring averages the pixel levels at a local scale, and when we subtract this average, any of the pixels that were less than their local average in the original image end up with negative values in the result. Since there are no negative values in image math, they all go to zero, leading to noise "holes."

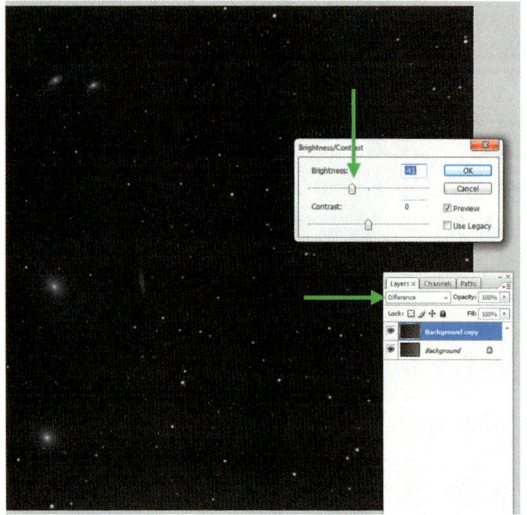

Figure 122. Setting a pedestal value and subtracting the map

We need to scale down our gradient map, subtracting only a fraction of it. The opacity slider is no help here, since that will only progressively blend in more of the original problem image. The Curves and Levels tools allow us to darken the gradient map, but not proportionally at all levels. The tool to use here is the Brightness/Contrast tool (Image > Adjustments > Brightness/Contrast). The brightness slider simply multiplies the RGB value at each pixel by the same value. Sliding it to the left divides each value. Adjust the slider to the left until the background returns to a more normal level. This is shown in Figure 122.

Repairing satellite trails, dust spots, and reflections

Sometimes, despite our best efforts, satellite or airplane trails slip through calibration to the final image. While it's best to eliminate them in calibration with sigma stacking methods, sometimes that's either not appropriate (when there are few exposures) or the trails survive stacking anyway. If the images were taken with a monochrome CCD through filters, the trails will be on separate color channels. Any removal actions should only be applied to the relevant channel in this case.

Adobe introduced "content-aware" tools with Photoshop version CS5. These tools do an excellent job mimicking not only the colors, but also the textures of their surroundings. This simplifies a lot of cosmetic repair tasks. A quick way to repair trails with these newer tools is to simply select the trail as closely as possible with polygonal lasso tool, then select Edit > Fill (or Shift-F5). Make sure the "Content Aware" option is chosen in the Content drop-box.

In earlier versions of Photoshop, it can require a little more effort. One method is to use the Spot Healing Brush. Zoom in on the trail and use the narrowest brush you can. Try the Proximity Match setting for the tool if the results from the Create Texture setting look too artificial. Work in small segments so you can evaluate the results as you go. Another option to consider is the Patch tool, which works similarly to the Proximity Match setting of the Spot Healing Brush. Use the polygonal lasso to closely select the trail, and then choose the Patch tool from the toolbar. Click within the selection and move the cursor around while still holding the mouse button down. This tool allows you to replace the selected area with those in a region under the cursor of the same shape. Usually the area just to one side of the selection will be similar enough to substitute, especially if the polygonal selection is very narrow.

With these methods, pay close attention to any bright stars or objects the repair passes through. The default repair may be unsatisfactory. For bright stars, you can either not correct the trail where it intersects the star, or you can copy a similar star, or piece of a star, from elsewhere in the image.

Correcting elongated stars

Small errors in tracking, wind, poor polar alignment, or optical distortions can produce oblong stars. While any such shift or distortion produces a blur across all affected areas, it is particularly noticeable on stars because of their high contrast. The effect on an extended object cannot be simply reversed, but at least the stars can be reshaped.

Figure 123. Oblong stars

The simplest Photoshop trick for minor reshaping of stars is to combine the Move tool with the Darken blending mode. Figure 123 shows a small area of a problem image. These stars are sitting on top of background nebulosity. To correct the stars, copy the image onto a new layer, and set the blending mode to Darken. Recall that the Darken mode will compare each pixel in this layer and the image below it and reveal the darker of the two. With the same image on top, there is no change, but if we shift the top layer a few pixels in the direction of the stars' distortion, only the center of each star will overlap, and the overhanging ends will not show through as long as one of the layers has background there. To shift the layer, either use the Move tool (the first tool on the tool bar) or the Offset filter. We're only shifting by a few pixels, so use the arrow keys if you use the Move tool.

Figure 124 shows the result, but it also illustrates why this sort of correction is far from ideal. On a very dark background, this method can be very effective, but on nebulosity it tends to leave artifacts. At a distance, these small details won't be noticed in the image, and no doubt it looks better than before, but detail has clearly been lost. Change the opacity of the layer to fade the effect. For a little more control, you can select only the stars (with the color range tool, or any other method), feather the selection, and then offset the selection. This will limit the effect to the stars, but any such correction is only a last resort. It's always worth the time to prevent the underlying equipment issues whenever possible.

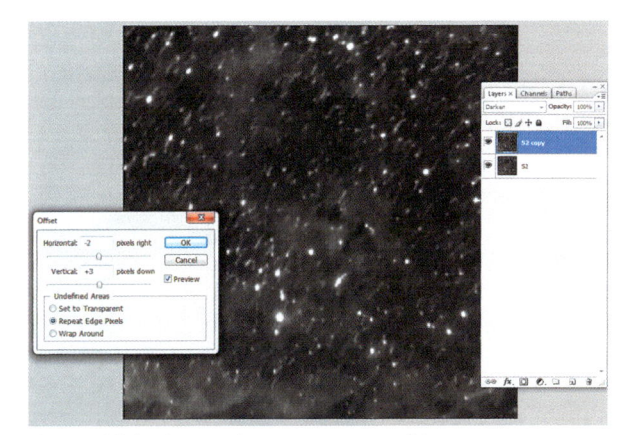

Figure 124. Limited correction to oblong stars

19 Color synthesis and adjustment

If you image with a DSLR or one-shot color camera, your images arrive at the computer in full color. But if your camera is monochromatic, you must synthesize your filtered images into a color image. To create a color image with monochrome data, we map a separate grayscale image to each color channel. The relative brightness in each channel determines the color. These grayscale images may have been taken through standard red, green, and blue filters to create a true-color image, or they may be narrowband exposures that will create a false-color image. Whatever their source, optimize each image before combining them. It is best to correct gradients, fix cosmetic defects, and reduce noise separately for each channel, then bring them together.

RGB Color Synthesis

In order to combine our red, green, and blue filtered grayscale images into one color image, we assign each to the corresponding channel of an RGB image. With the three filtered grayscale images open, you can copy and paste each one into the appropriate color channel (Figure 125) in the Channels tab.

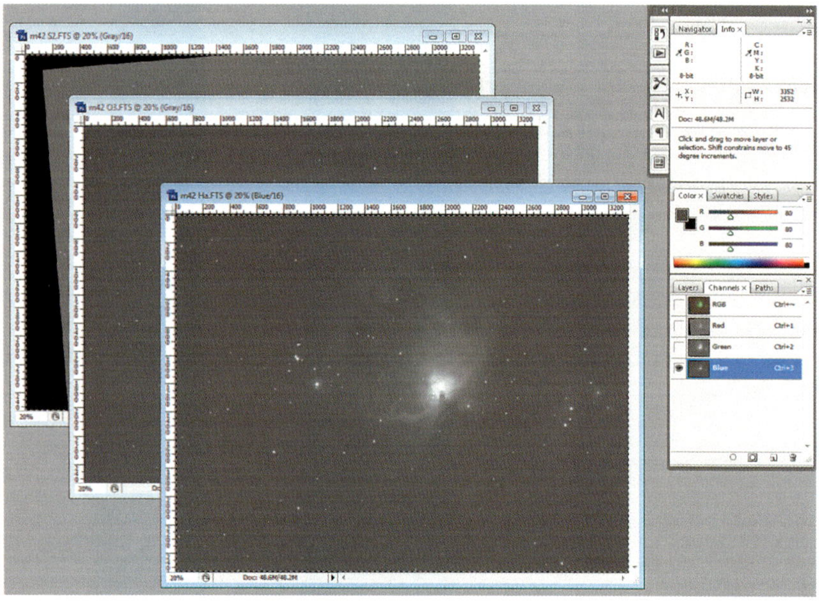

Be sure that the image into which you are pasting is set to RGB mode, otherwise you will only see one channel. Usually, it's simplest to change the mode of one of the three grayscale images to RGB, and then paste the other two into the appropriate channels.

Alternately, Photoshop has a Color Merge dialog that facilitates this process (Figure 126). This dialog is found on the Channels panel via the little menu expansion icon in the upper right. Note that the Color Merge option will be inactive ("grayed out," ironically) unless there is at least one other grayscale image of the same size open. The current image must also have only a single layer.

Creating a luminosity channel

For our standard RGB image, there are only three channels: red, green, and blue. You've heard that LRGB is the way to go for low-noise images, so you've also taken unfiltered luminance exposures, but Photoshop doesn't seem to have an LRGB mode. How are we to integrate the calibrated luminance image? There are two ways to do this.

The first is to add a new layer to our existing RBG image, and paste the luminance image into that layer. Then set the layer blending mode to Luminosity. This is approach has the benefit that we can adjust the impact of this luminosity layer via the layer opacity slider. When set at 100% opacity however, it can wash out the color saturation.

Another approach is to utilize the image mode that has a true luminance channel: LAB. Once you have created the RGB image, simply change the image mode to Lab Color (Image > Mode > Lab Color), paste the luminance image into the luminance channel, and change the mode back to RGB. This method tends to produce richer colors while retaining the noise-reducing benefits of adding a luminance channel. Once you are back in RGB, the

Figure 125. Pasting images into each color channel

Processing Images

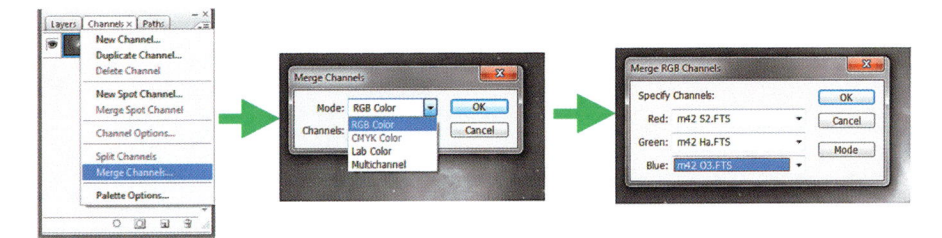

Figure 126. Using the Merge Channels option

effect is not as easily attenuated as the blending mode method, however.

Separating the luminance and color data is important for many processing techniques. The luminance data typically contains finer details, as colors change over larger spatial scales than changes in brightness. This concept of separating luminance from chrominance will come up again and again as we review various processing methods. If you didn't capture separate luminance subexposures, you can create a synthetic luminance channel. And if you do have luminance exposures, you can further improve the SNR of this data with synthetic luminance data. The goal of a synthetic luminance channel is to take full advantage of all available exposures to increase the SNR.

Without separate luminance exposures, you can create a synthetic luminance image by combining the red, green, and blue images together into one grayscale image. This will capture the SNR benefit of the total exposure time. A synthetic luminance image can be created either at the stacking stage or by working with the image in post-processing. To create a synthetic R+G+B luminance at the stacking stage, stack all of the red, green, and blue subexposures together—in groups with their own flats—using a statistical method that will account for their inherent differences in brightness. Sigma combining methods do not handle these differences well, as they will exclude a large fraction of the exposures. A simple mean method will do if there aren't too many cosmetic issues to correct (satellite trails, etc). Otherwise, an entropy-weighted or HDR method can be used, but verify that the noise is lower than in the original image.

The second and simpler way to create a synthetic luminance channel is in post-processing. This is especially useful if the image data came from a one-shot color sensor. Duplicate the current image (Image > Duplicate), then either change the image mode to grayscale (Image > Mode > Grayscale) or use the Channel Mixer tool (Image > Adjustments > Channel Mixer) to assign specific weights to each color channel in the new image. The benefit of using the Channel Mixer is that if one channel has more noise than the others, you can add less of it to the final mix. Be sure to check the "Monochrome" box to produce a monochrome output if you use this tool. Now change the mode of the original color image to Lab Color and select the L channel on the Channels window. Copy the entire grayscale luminance image (Ctrl-A, then Ctrl-C) and paste it into the L channel of the original image. You can make any luminance-specific adjustments (sharpening, high-pass, curves, etc) to this channel now or change the mode back to RGB and make them later.

If the luminance and RGB exposures were captured at the same resolution, you can go a step further and combine the true luminance images with the R+G+B synthetic luminance to create a L+R+G+B "super luminance" channel that captures the entire exposure time.

Narrowband RGB synthesis

Nearly all color images created with narrowband data are false-color. The three primary emission lines we capture do not provide much in the way spectral diversity: H-alpha and SII are both deep red, and OIII is blue-green. Further, H-alpha typically outshines the other emission lines by an order of magnitude for most objects. Thus, a full-spectrum color image of most emission objects simply looks red. If we want natural colors, it's easier to take a regular RGB image than bother with narrowband filters. The real fun—and some would argue, departure from reality—comes when we map the narrowband data from each ionization line to more distinct color channels. This emphasizes the differences between areas where different gases dominate, so while both H-alpha and SII are red, we can visually separate them by making one red and the other green in a false-color image. We can also stretch each color channel so they are of similar brightness to create better color balance. Revealing these gaseous variations in nebulae allows us to create images that are both dramatic and scientifically accurate where a "natural" color mapping would be drab.

The choice of what emission lines to capture and to what colors they are mapped is arbitrary. NII and H-beta filters

141

Figure 127. IC 405 and 410 rendered in Hubble (left) and CFHT (right) palettes

are sometimes used, usually for planetary nebulae, but H-alpha, SII, and OIII are by far the most popular filters. The most common mapping is the wavelength-ordered palette, also known as the HST (Hubble Space Telescope) palette: red = SII, green = H-alpha, blue = OIII. Many of the most famous deep-sky images use this palette. Another way to arrange narrowband colors is the CFHT (Canada France Hawaii Telescope) palette: red = H-alpha, green = OIII, blue = SII. Either of these are a starting point for color mapping. Channels are usually blended and the hues can be shifted further in subsequent processing.

Figure 127 shows the same narrowband image of IC 405 and IC 410 rendered in both the HST and CFHT palettes without any channel blending or additional hue adjustments.

> Creating a synthetic luminance with narrowband data is more challenging than with RGB data. Because the OIII and SII signals are typically far weaker than the H-alpha+NII signal, the best approach is to stretch each image before combining them. Creating a synthetic luminance in calibration without stretching the OIII and SII data first will result in the H-alpha dominance we try to avoid.

Further hue adjustments

An image rendered in the Hubble palette may look bland at first. The colors seem to range only from the middle of the spectrum from yellow to cyan, with no true reds or blues. This is due to the relative strength of H-alpha in most objects and the fact that these regions typically overlap with the other ionized gases. Even curve adjustments to each channel don't seem to create much color contrast. Once most of the post-processing is done, the hues can be adjusted to emphasize the existing color contrast and broaden the palette to include the full spectrum. These adjustments are a matter of personal preference and experimentation.

To achieve this effect, start with an HST palette image and use Curves to tune the individual color channels so the green of H-alpha doesn't dominate. Starting with a range of colors is critical to any subsequent hue adjustments. Then use the Selective Color tool (Image > Adjustments > Selective Color, or via an adjustment layer) to modify the existing image. It helps to apply this as an adjustment layer, since you may want to change the settings later. There are many possible hue modifications, but one way to expand the spectrum of an HST palette image is to:

- **Shift the greens (areas of mostly H-alpha) toward yellow.** Choose "Greens" from the Selective Color tool's drop-down box and use the sliders to remove most or of the cyan and some of the magenta. Depending on the original hue and your taste, it may take a little or a lot of magenta removal to create a nice yellow.
- **Shift the yellows (areas where there is a mix of SII and H-alpha) toward red or orange.** Choose the yellows from the drop-down and remove most or all cyan, but adjust the magenta slider until they take on an orange-gold tone.
- **Shift the cyans (a mix of OIII and H-alpha) toward blue.** Choose the cyans from the drop-down and remove most or all yellow and again adjust the magenta until they are the shade of blue or turquoise you prefer.
- **Adjust the blues to taste.** An adjustment to the blues' lightness via the black slider is useful, since the OIII channel is frequently the dimmest. The blues can also be adjusted similarly to the cyans, depending on the hue you want to achieve.

If any of these hue adjustments dramatically altered the brightness of an area in an undesirable way, the black slider can be used to correct any luminosity shifts. Figure 128 shows the same IC 405 and IC 410 HST palette image with a quick application of these hue adjustments for comparison.

Figure 128. An example of the Hubble palette with hue adjustments

These adjustments are shown to illustrate one possibility, albeit a popular one. Each image will call for its own palette. These are artistic choices, and there are infinite combinations available for every imager to explore.

Using clipping layer masks to map color

Dr. Travis Rector, a professional astronomer at the University of Alaska, has described a more flexible method of color mixing in Photoshop that allows you to assign monochromatic images to any specific color with an unlimited number of potential color channels. This method has become a very popular technique among narrowband imagers.

In Adobe parlance, a "clipping" layer mask is one whose effects are only applied to the layer below. This is noted visually with a bent arrow icon on the layer panel (Figure 129). This technique uses clipping layer masks applied to a Hue/Saturation adjustment layer to colorize a monochrome layer below it. When all of the colorized layers are set to the Screen blending mode, the effect is that they combine into a full-color image. Recall that the Screen mode mim-

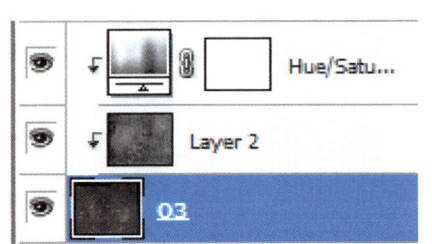

Figure 129. Clipping layers are indicated by a bent arrow

ics the additive effect of multiple projectors shining on one screen. This allows us to assign a monochrome image to any color we wish, building up a color image without being limited to composing it from RGB or CMY trios. We can even use more or fewer color channels, depending on the data we have.

To synthesize a color image with clipping mask layers:

1. As with standard color synthesis, the grayscale images should already be stretched to fill most of the dynamic range, cosmetic defects should be repaired, and noise reduction and contrast enhancements done.

2. Create a new RGB image of the same size as the aligned grayscale images. Paste each grayscale image into a new layer in the RGB image, and label each appropriately.

3. Add a Hue/Saturation adjustment layer to each monochrome layer, and check the Colorize box for each. The color is chosen as an angle on the color wheel, with 0° representing pure red, 120° for green, and 240° for blue. (Or if you prefer, 60° for yellow, 180° for cyan, and 300° for magenta.) Select the appropriate color angle for each adjustment layer, with the saturation set to 100. The angle does not have to represent a pure color—it can be any arbitrary color, which is the advantage of this method. The lightness can be set depending on how many channels you are combining. A good starting point when combining three channels is -66 (one-third original brightness for each), but this can be adjusted later.

4. Now, link each adjustment layer's effect to only the layer below it by activating a clipping mask (Layer > Create Clipping Mask or Ctrl-Alt-G). Then set the blending mode for each of the grayscale layers to Screen. The Screen mode is analogous to having a separate projector shining each layer onto a screen, building up color and brightness.

The image should now be a mapped color image. Each color channel can be individually adjusted by adjusting the lightness, saturation, and color angle of each layer. The underlying monochrome data in each channel can also be adjusted separately, either by working directly on the layer or through additional adjustment layers.

Blending RGB and narrowband data

One common issue with narrowband images is star color. Balancing the channels to produce a pleasing nebular image can lead to unnatural star colors like magenta or green, and even when the stars look reasonable, the colors do not represent their actual hue. The only way to get true star colors in a narrowband image is to shoot the stars separately, either with RGB filters or with a one-shot color sensor.

This image can be added as a layer on the final narrowband image, with the Color blending mode selected. The layer must be registered and aligned with the narrowband image, and it should be masked to reveal only the brighter stars. This is easily accomplished with the Selective Color tool, though other methods are covered later. Use the Selective Color tool's additive eyedropper (the one with the '+' next to it) to select the stars you want to include, and then save the selection so you can revert to it if needed later. Expand the selection a few pixels, and feather it about half as much as the expansion. Create a layer mask from this selection (Layer > Add Layer Mask > Reveal Selection). If the brightness of the color image is dramatically different from the narrowband, try the Hue blending mode or adjust it via the Curves tool. It may take some experimentation with the selection to get the mask right. Color saturation can be adjusted directly to the color layer or via an adjustment layer.

Narrowband data can also be blended into an RGB image to enhance the detail and luminosity of emission nebulae. In this case we want to affect luminosity but not color, so the narrowband data should be applied as a layer to each color channel in Lighten mode. This is straightforward if your color channels are in separate layers via the clipping layer mask method above, but if they are not, you can always copy a color channel's monochrome data to a separate image, adjust it, then copy it back via the Channels panel.

H-alpha data works very well when applied to the red channel. If the effect is too strong, adjust the opacity slider. SII is also red, but since the H-alpha signal is typically orders of magnitude brighter, there is rarely a reason to apply it to an RGB image. OIII is blue-green, so apply it to both the green and blue channels. Many imagers also apply a small amount of H-alpha to the blue channel, depending on the object. This can help create the "electric pink" color of H-alpha in most color images.

Color balance

As you stretch an image, small imbalances in color can be exaggerated. Use the histogram window to keep an eye on the each color channel throughout processing. As covered in Chapter 17, the Color Sampler Tool is very helpful to monitor the exact RGB values for a few spots of the background. In general the background should be a neutral gray, so each color value should be very similar for a balanced image. Differences between points spread around the image can also help reveal subtle color gradients.

The Color Balance tool (Image > Adjustments > Color Balance) is a simple tool to adjust the overall hue of an image. If a near-final image needs a small tweak, this is an easy way to make the correction. A better way is to use the Curves tool to directly adjust each color channel. This allows specific modifications to the levels that are out of line—sometimes the background can be balanced while the midtones are not.

The bluntest tool of all is Photoshop's Auto Color tool (Image > Adjustments > Auto Color). It has no parameters to adjust; it automatically adjusts color balance and saturation. It would seem to have no place in an astronomical imaging workflow, but once your image is nearly finalized, give it a try. In some cases it can produce surprisingly good results, which can be faded (Edit > Fade) or completely undone if they don't work.

Enhancing color saturation in RGB

We want our images to be colorful despite the fact that the universe typically offers us something fairly drab. Color saturation can be improved in several ways, and at multiple stages of processing. We've seen that stretching an image can desaturate colors, so it helps to apply periodic saturation boosts during that process. Saturation is also adjusted near the end of post-processing, where it is a key factor in creating an attention-getting result. But no matter when or how much you punch up the saturation, remember to apply it selectively.

Increasing saturation in the background will exaggerate the natural chrominance noise (usually showing as color speckles) there, so be careful to apply most saturation adjustments to the object or areas of interest. This is especially true of one-shot color images and uncooled sensors. The more noise there is to begin with, the more it will stand out as you stretch and apply saturation adjustments. Isolating the richest color saturation to the main subject also enhances the perception of depth, making the subject "pop" in the final image.

The most obvious tool for increasing saturation is the Hue/Saturation tool, but aggressive use of this tool can add chrominance noise. Saturation can also be increased via the Match Color tool (Image > Adjustments > Match Color), matching the image to itself, then using the color intensity slider to increase saturation. This seems to add less color noise than other methods, but the hack nature of using a tool designed for another purpose speaks to the fact that Photoshop's tools do not usually have astronomical images in mind.

Boosting saturation in LAB

While the Hue/Saturation tool works well in many circumstances, there is another way to increase saturation that provides more precision and less noise. As covered earlier, LAB mode breaks an image into a luminosity channel and two color coordinate channels called A and B. While the A and B color channels are not as conceptually intuitive as RGB, they give us the power to increase color saturation completely independently of luminosity.

The basic method for increasing saturation with LAB mode is simple. As an example, we'll boost the color in the relatively drab image of M31 shown in Figure 130.

Figure 130. Our initial M31 image

First, make a copy of the layer you want to adjust. Then change to Photoshop's Lab mode (Image > Mode > Lab Color). Photoshop will warn you that changing modes can affect the appearance layers, but choose "Don't Flatten." Unless you are using Adobe's Smart Object layers, you can keep your layers intact without worry.

Now that you're in LAB mode, the channels display will show Lightness, a, and b, instead of Red, Green, and Blue. Open up the Curves tool and change the tools operating channel to "a," as in Figure 131.

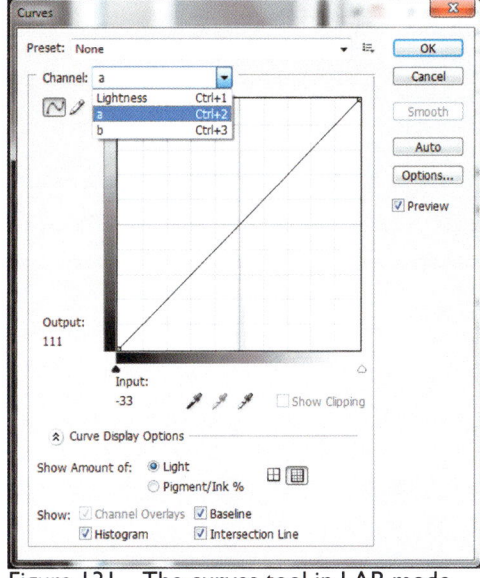

Figure 131. The curves tool in LAB mode

Our image has a relatively limited color palette, so all of the color information for both the a and b channels lies in a relatively narrow spike. To increase saturation, we are going to expand the dynamic range of these spikes. Just as increasing the slope in RGB mode expanded the dynamic range, resulting in a wider histogram where the slope was greater than 45°, here we are stretching the color information so that a narrow range of colors are mapped to a wider range. By doing this to both channels equally, we get an increase in saturation without altering hues. If you prefer, these transformations could just as easily be done with the black and white point sliders in the Levels tool.

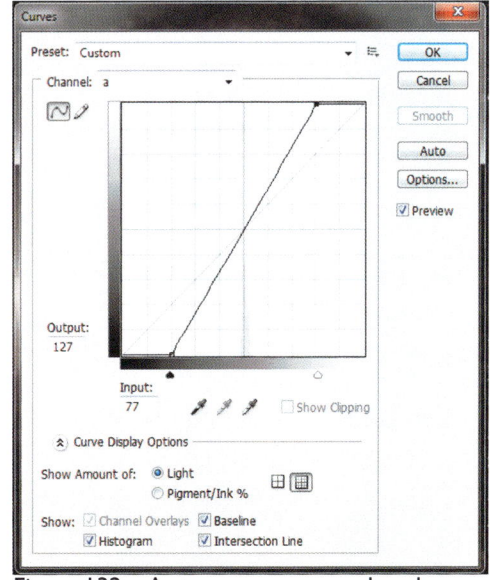

Figure 132. A steep curve expands color

Figure 133. The results of a saturation adjustment via LAB color

In order to avoid shifting the hues of the image, it is important to make symmetrical adjustments to the black point anchor and the white point anchor. In other words, your straight line must cross the middle of the histogram. Here, inputs of 77 on both sides were mapped to 127.

Now, repeat the exact same straight line curve for the B channel, with the same values that were used on the A channel. After applying identical curves to both the A and B channels, the histogram shows a modest expansion of these color channels with no change to the luminance channel. This small change in the histograms of the color channels creates a substantial increase in color saturation. Note that the luminance histogram is unchanged in Figure 134. The final image is shown in Figure 133.

To blend the results into the image, change the newly saturated layer's blend mode to Saturation. This will only bring in the color saturation information from the layer. Then use the Opacity or Fill slider to adjust how much of the new saturation to apply to the image.

If you want to apply the saturation to a portion of the image, add a layer mask (set to hide the top layer) and paint white over the areas you wish to be saturated. Be sure to use a feathered brush or blur your painted area in the layer mask. Once you are done with LAB mode manipulations, you can change the image mode back to RGB.

> To quickly add saturation selectively to specific areas of the image, try Photoshop's Sponge tool, located in the tool bar under the Dodge and Burn tools. With the mode set to Saturate, you can selectively enhance saturation via the painbrush. This tool is particularly useful for restoring color to large, blown-out stars by bringing in some color from the halo.

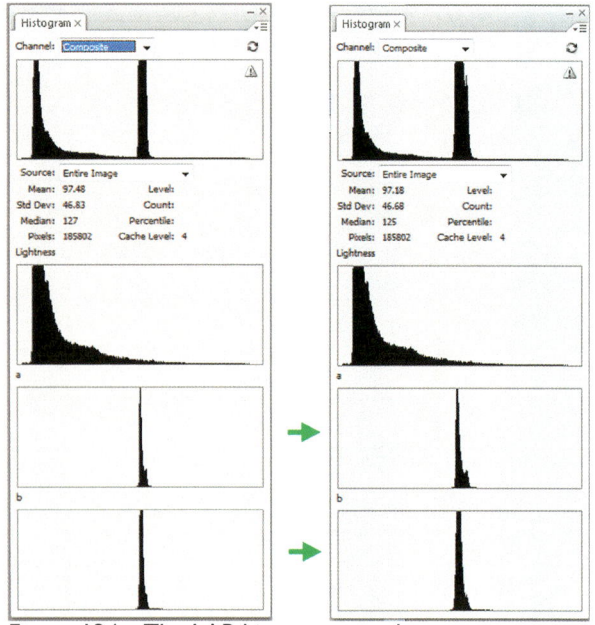

Figure 134. The LAB histogram results

20 Sharpening and local contrast enhancement

The effect of most sharpening algorithms is to emphasize edges. Boundaries of different brightness are emphasized by making the darks darker and the lights lighter. Unlike a contrast-enhancing curve however, we can determine the scale at which sharpening tools are applied. With a small "radius" setting, tiny transitions are enhanced, while larger scale changes are ignored. As we increase the radius, the small details are less affected and contrast is created across broader structure. Most sharpening tools also have a strength setting to attenuate the effect.

Photoshop and other image processing packages offer a variety of sharpening tools, but it is usually a mistake to apply them to the whole image. Sharpening algorithms have no way to distinguish between small details that are real and those that are simply variations in brightness caused by noise, so if the signal-to-noise ratio is low, sharpening can enhance noise. Think of sharpening and noise reduction as opposite functions. With noise reduction, we want to smooth out the dimmest areas where SNR is low. We want to apply sharpening to the brighter areas where SNR is high.

Over-sharpening can create image artifacts, especially where there are severe transitions like stars and hot pixels. If the strength is high and the radius is too large, it can lead to halos around stars, an effect known as "ringing." Since stars typically have abrupt boundaries with the background, sharpening also has the unfortunate consequence of increasing their prominence. Separating stars from the rest of the image before using any sharpening filters can help.

When used for daylight photography, sharpening is applied at a small scale (alternately expressed as a high spatial frequency) because these images are rich in small details to be revealed. Using a similar approach in astronomical images will look unnatural because deep-sky objects are mostly diffuse. Rather than defining small edges, we want to create contrast on a larger scale to highlight features like galaxy lanes and nebular structure. Adjustments like this create local contrast enhancement that is tuned to the characteristics of the image rather than just enhancing small edges.

A fundamental principal of astronomical imaging is that luminance changes at a higher frequency than chrominance—in other words, scanning across an image, color changes much more slowly than brightness. In order to avoid creating color artifacts and keep chrominance noise low, isolate the luminance and apply sharpening techniques only there. This can easily be done by working with the LAB color mode's luminance channel or through a layer set to the Luminosity blending mode. Any of the methods that follow can be applied this way, depending on the content of your image.

Convolution, deconvolution, and sharpening

While a sharp image can never be completely recovered from a blurry one, we can in theory reverse the effect of specific optical distortions. We've seen that a point source of light like a star is spread into a specific pattern, the Airy disk, as a result of diffraction. Additional distortions are added by your optical system and by the atmosphere. These transformations are known mathematically as a point spread function (PSF), and the translation from the original state to the blurred one is called convolution. Image processing filters that blur an image do the same thing—they apply a discrete point-spread function, known as a kernel, sequentially to every pixel in an image. The kernel determines each pixel's new value by incorporating a specific amount of each of the surrounding pixels' values, and this averaging process causes the blurring. A large kernel incorporates information from pixels that are more distant from the original, resulting in greater blurring.

Deconvolution aims to reverse the specific effects of convolution by applying the inverse point spread function to the one that caused the blur. The convolution process is destructive, so applying the inverse function does not completely restore the original image, but the better we can estimate the PSF, the more of the original detail we can restore. Most sharpening filters, including unsharp mask, simply apply kernels of various designs to an image. The closer the filter's kernel is to the inverse of the optical system's PSF, the closer the result will be to the maximum possible restoration. Since we can never perfectly characterize the PSF, the best estimate is typically a Gaussian-shaped kernel, with the size approximately matching the atmospheric seeing. It is hard to give specific guidance on appropriate deconvolution or sharpening settings, since they are dependent on the optics and atmospheric conditions,

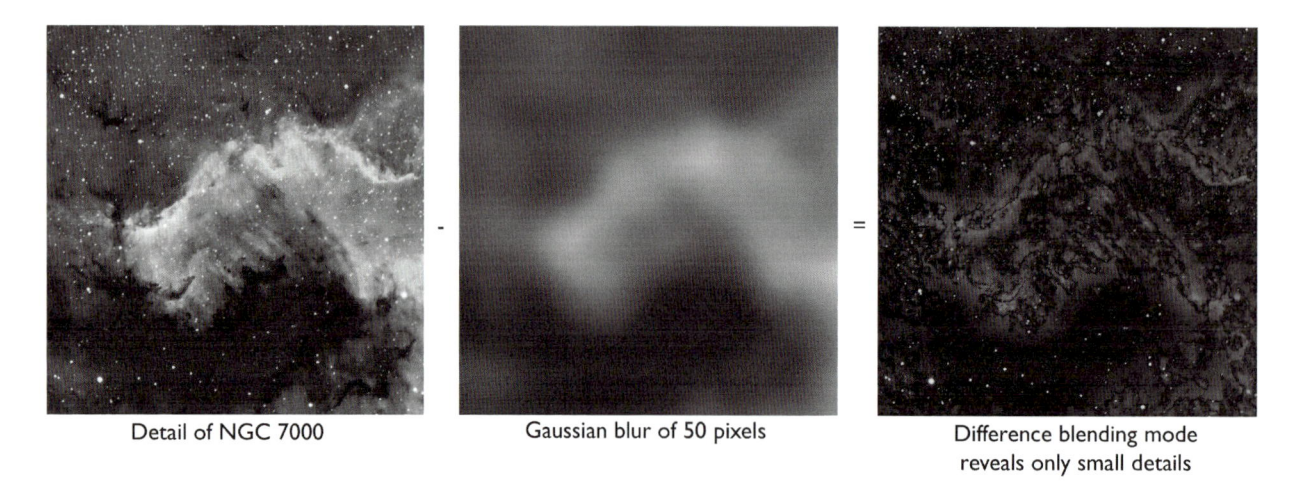

Detail of NGC 7000 − Gaussian blur of 50 pixels = Difference blending mode reveals only small details

Figure 135. Subtracting a blurred image reveals small details

as well as the structure of the deep-sky object and the details that you wish to highlight.

When we speak of true deconvolution, we are referring to something more complex than a simple sharpening algorithm. Many deconvolution methods have been developed, and they iteratively attempt to recreate the original image through a step-wise process of blurring the existing image with a specific PSF (which can be estimated from a star), then adding back to it the difference between itself and the blurred version. The resulting image is then put through the same steps. The most commonly used deconvolution method is the Richardson-Lucy algorithm, which is implemented in a number of scientific image processing programs including IRIS, CCDStack, ImagesPlus, MaximDL, and PixInsight. Photoshop does not currently have any deconvolution capabilities, though there are commercial plug-ins available that implement it. Deconvolution is a processor-intensive task, even for modern computers, so it can take minutes or even hours to run a few iterations.

Unsharp mask and local contrast enhancement

Generally, the only sharpening filter in Photoshop to use on astronomical images is unsharp mask. The unsharp mask filter has a funny name that seems to imply blurring, but it is one of the most powerful tools for sharpening and for contrast enhancement. You can target the scale of structures where contrast will be enhanced with the radius setting. By using a large radius value, you can create contrast between large structures in the image that would otherwise be nearly impossible with Curves or other tools. Daylight photographers have used this trick since the film era to bring out detail that otherwise looks muddy or hazy, and it works as dramatically in astronomical images.

The name unsharp mask comes from the original film negative method, which paradoxically uses a blurred image to create greater contrast for small details. When printing the negative, a positive of the image (the mask) was placed in front of it, with a glass plate between the two to put the positive image slightly out of focus (unsharp). This blurring meant that only the larger structures would be can-

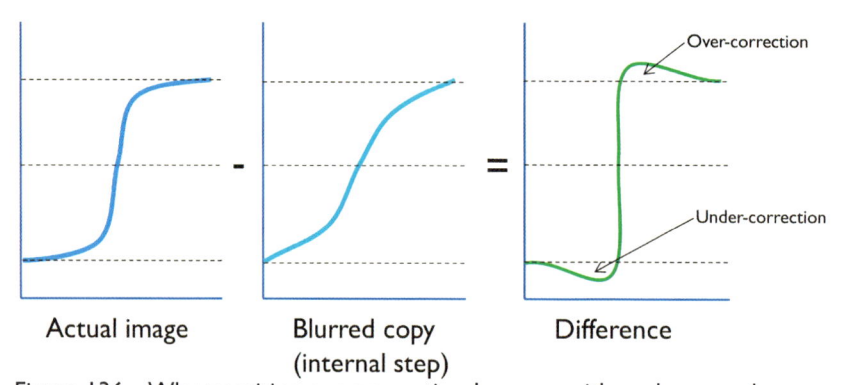

Actual image − Blurred copy (internal step) = Difference (Over-correction, Under-correction)

Figure 136. Why transition overcorrection happens with unsharp mask

148

Processing Images

celed by the positive, sparing the details. The same thing can be accomplished manually in Photoshop. To reveal only the high-frequency details in an image, you can make a blurred copy on a new layer and set the layer blending mode to Difference (Figure 135). Adding the resulting image back to the original will accomplish the same thing as the unsharp mask filter, though the filter is obviously faster.

Unsharp mask is not without its drawbacks, however. Figure 137 shows a hypothetical perfect view on the left, with the blurred view we capture after diffraction, the atmosphere, and optical aberrations shown in the middle. On the right is a slightly overzealous application of unsharp mask, applied to the middle image in an attempt to recover the left image. While the result is an improvement, the filter over- and under-shoots at transitions, creating halo or ringing artifacts. Figure 136 shows why this happens. The blue line shows the brightness levels at a transition, and the cyan line shows the blurred version of this transition used internally by the filter as a mask. (The filter actually works through a kernel method, but the effect is the same.) When the blurred mask is subtracted from the original, the result is a transition that is sharper, which you can see by the steeper vertical slope in the middle, but that overcorrects initially on one side and undercorrects on the other. In astronomical images, this effect is most pronounced around stars when the radius is set close to that of the stars.

The best way to use sharpening or contrast enhancement is to apply it to the luminance channel, especially if small spatial scales are targeted. Any existing chrominance noise will be enhanced. Since the luminance values in an image change brightness over shorter distances—a higher spatial frequency—than color, limiting sharpening to luminance allows us to enhance those changes without affecting the color data where it wouldn't be appropriate. Figure 138 shows a close-up of how existing color noise is made worse when sharpening is applied to a color image, and how sharpening only the luminance information mitigates this effect.

Unsharp masking will also increase the overall brightness and size of stars, and it can make stars seem more like uniform disks by creating overly abrupt transitions at their edges instead of a natural Gaussian shape. Because of this, many imagers prefer to separate the stars to a separate layer be separated before sharpening. Methods for doing this will be covered later.

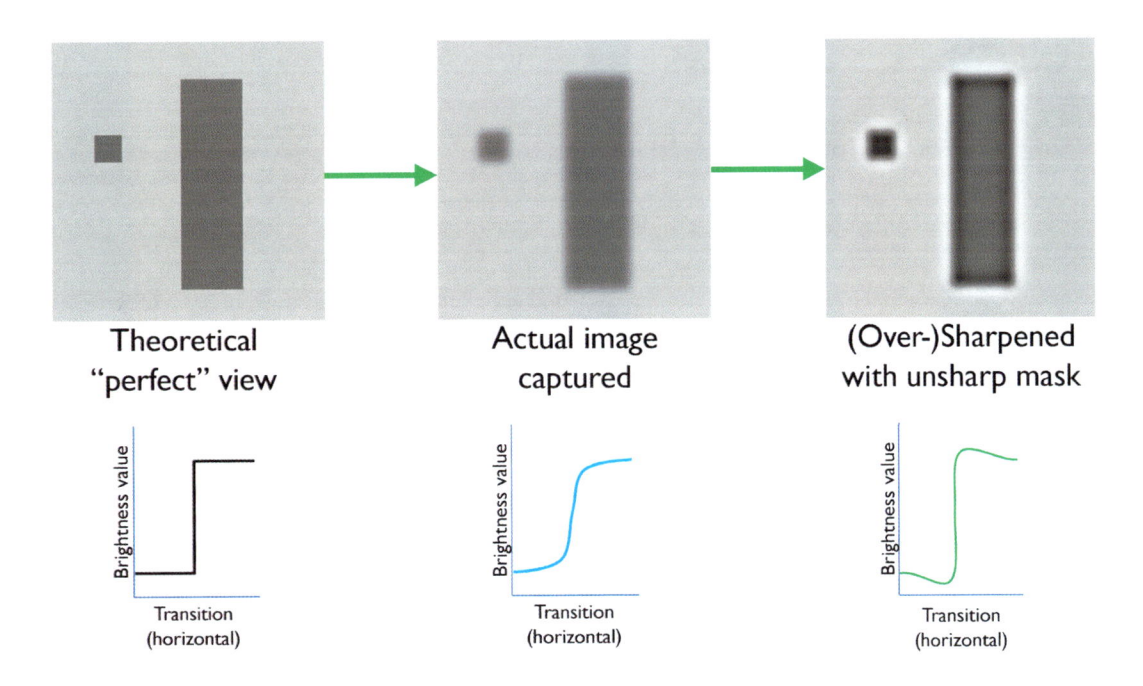

Figure 137. Sharpening artifacts due to under- and overcorrection

Figure 138. Reduce color noise by applying unsharp mask only to luminance

Figure 139 shows a region known as the "Elephant Trunk" from an image of IC 1396 where the unsharp mask filter has been applied with varying radii. In each case, the filter was applied with settings of 125% for strength and a threshold of 2 levels. At the top is the original image. In the enlarged crop (on the right) from the image with a two-pixel radius, you can see that while small details are clearer, the noise has also been emphasized. The edges around the stars are also starting to show some haloing. At five pixels, the stars have significant halos, but the contrast across the image is also stronger. Neither of these setting would probably be appropriate for this image, though small radii like this could be used with a greater threshold setting to prevent halos.

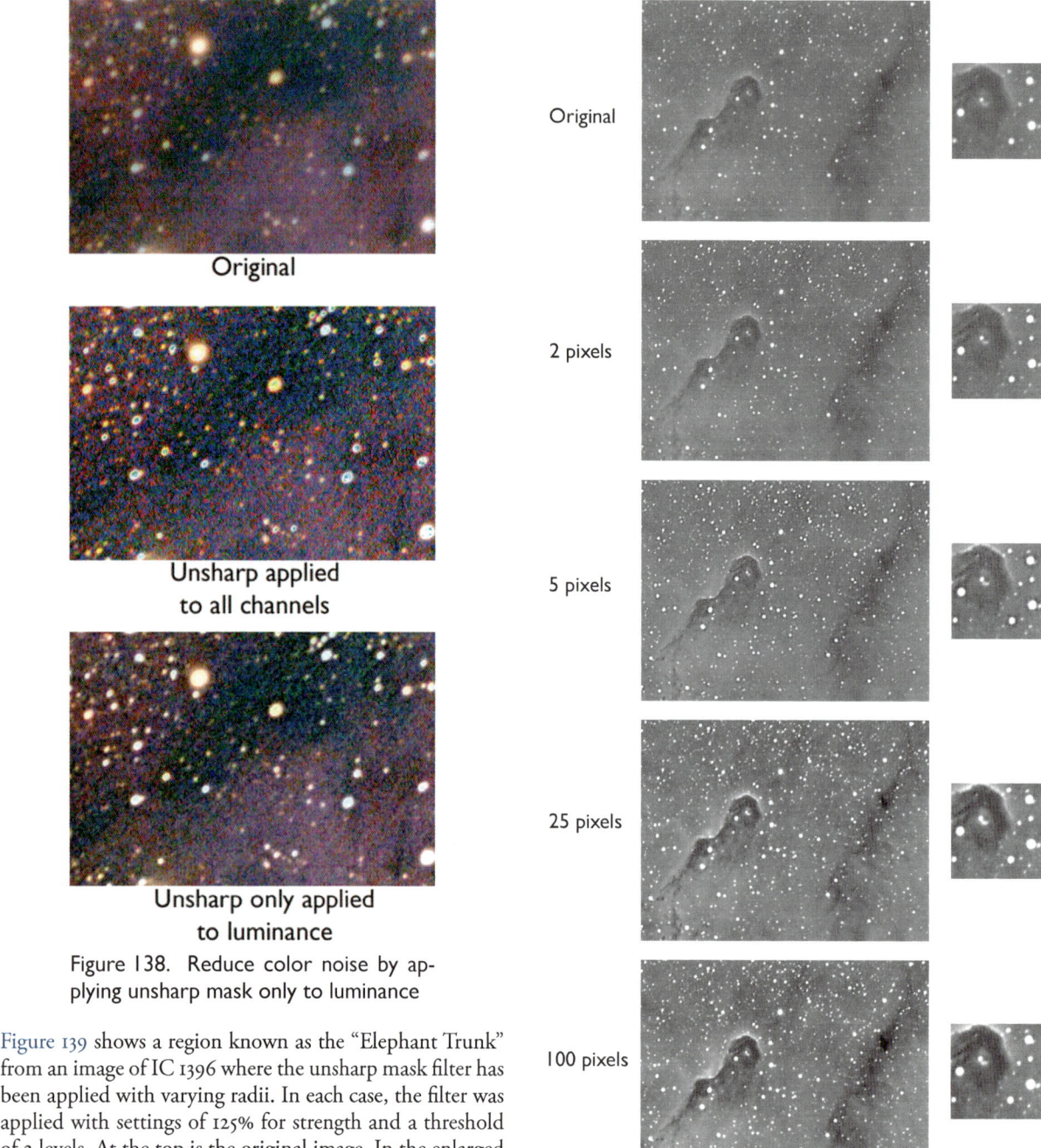

Figure 139. The effect of varying the radius of the unsharp mask

Increasing the radius to 25 pixels, the star halos have disappeared, because those transitions are small enough to fit entirely within the kernel's radius. Larger structures in the image now have substantially greater contrast—the Elephant Trunk stands out and vertical striations in the nebulosity are starting to show up above it. With some noise

Processing Images

reduction and other minor adjustments, this image could be used as a luminosity layer in a final image.

At a radius setting of 100 pixels, there are only a few differences from a setting of 25, since we are probably exceeding the size of the structures we want to emphasize at this point. The dust lane at right is stronger here, but overall the effect is simply greater contrast everywhere. In an image with greater resolution, this may be a more appropriate setting, but in this case it is probably overdoing it.

Selective application with layer masks

You can also go a step further and proportionally apply sharpening across an image based on the brightness level in the luminance channel, or even selectively to specific shades. To apply sharpening most to the brightest areas and least to the dimmest areas:

5. Copy the image to a new layer or make a composite layer to work with (Shift-Ctrl-Alt-E) and copy that. The idea is to have a sharp layer on top and the existing image on the bottom.

6. Sharpen the new top layer as needed. It can help to apply sharpening first at a larger scale, then again at a smaller scale. At this point it is okay to be fairly aggressive, since we'll fade the results via a mask.

7. Create that mask by clicking the layer mask icon on the Layers palette. Copy the whole image (Ctrl-A, then Ctrl-C). View the layer mask by holding Alt and clicking the mask thumbnail on the layers palette. Recall that white on a layer mask reveals the top layer and black hides it, showing the layers below.

8. The mask can be further adjusted via a Levels adjustment. Bringing down the white point expands the range of highlights where the sharpened image will be revealed. Likewise, bringing up the black point expands the range of shadows where the sharpened image will be hidden, and the midtone slider controls the slope of the transition between them.

Figure 140 shows an example image (of the nebulosity around Sadr in Cygnus) where a solid red layer is masked with the image itself to illustrate where the top layer is revealed. The original image is on top. A solid red top layer is added, then the bottom layer is pasted into the layer mask in the second image. The third and fourth images show the effects of using a Levels adjustment on the layer mask. Moving the black point limits the top layer's effect to only the brightest regions. Likewise, moving the white point broadens areas where the top layer is revealed.

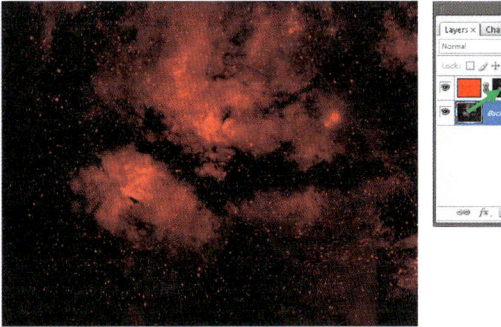

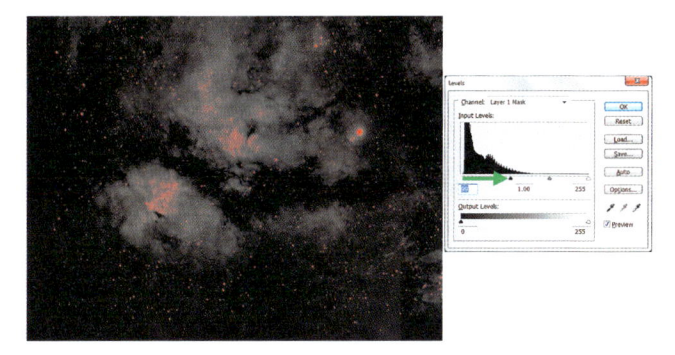

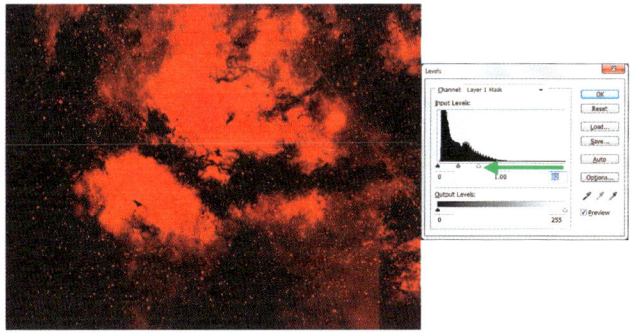

Figure 140. A solid red top layer shows the effect of pasting an image into a layer mask

The high pass filter

A high pass filter is similar to a sharpening filter, but instead of enhancing the high frequency details in an image, it retains only those details, rejecting everything else from the resulting image. Photoshop's implementation of the high pass filter reveals all of the edge transitions in the image smaller than the specified scale, and then covers everything else with a neutral 50% gray. This last part is important, because it allows us to combine it with the contrast family of layer blending modes, where 50% gray is the shade that leaves the underlying image unchanged. The effect is similar to what we saw before when recreating the unsharp mask filter by using the Difference blending mode. Figure 141 shows the same detail of NGC 7000 with high pass and unsharp mask filters applied using the same radius setting. Details at the same scale are enhanced, but the high pass filter shows the areas not affected in gray, while the unsharp mask filter retains the original image in those areas.

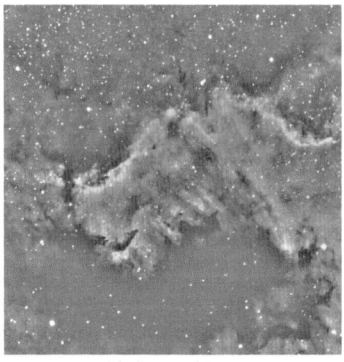

High pass filter
(50 pixel radius)

Unsharp mask
(50 pixel radius)

Figure 141. Comparison of the high pass and unsharp mask filters

As we've seen, the Overlay and Soft Light blending modes allow you to exaggerate color and luminosity—if both values are dark, the result gets darker; if both are light, the result gets lighter. Middle gray is unaffected. So pasting an image into a layer on top of itself and setting the blending mode to either of these results in an image with higher contrast. Since the high pass filter reveals only the edges and covers the rest of the image with gray, you can use it to selectively create contrast in parts of the image defined by the filter's radius.

The method is simple, and it can be done multiple times with different filter settings to emphasize details at different scales:

1. Duplicate the whole (nearly final) image into a new layer, with the layer mode set to Overlay or Soft Light. As usual a composite layer works well for this (Shift-Ctrl-Alt-E). Overlay mode is more aggressive, but the effect will be dialed back later through a layer mask.
2. Run the high pass filter (Filter > Other > High Pass) on the new layer. The radius setting controls the width of the frequency filter. Adjust the slider to create the desired effect on the areas you want to sharpen. At this point, it is okay to oversharpen, since we will limit the filter's effect in the next step.
3. Create a layer mask over the new layer that hides the filtered image (Layer > Layer Mask > Hide All or hold down Alt while clicking the layer mask icon on the layers panel.) The layer mask thumbnail in the layers panel will appear black.
4. Select the brush tool and make sure the current foreground color is white. With the layer mask selected, paint over the areas you wish to sharpen by revealing the high pass layer.
5. If the effect appears too strong, you have four options.
    » Reduce the opacity of the high pass layer.
    » Change the layer mode to Soft Light if it was set to Overlay.
    » If the details enhanced seem too large, start over with a smaller radius setting.
    » To control sharpening in specific areas of the image, paint over those areas with shades of gray in the layer mask to reveal less of the high pass layer.
6. If the effect is too weak, try stacking another high pass layer on top of the existing one. (Right-click and select Duplicate Layer.)

Other ways to create local contrast

Layer blending modes. As we've seen, several of the layer blending modes can be very useful for astronomical images. In particular the contrast family of layer blend modes can be used to enhance the contrast of an image. This method works well when selectively applied via layer masking. The idea is to create a new top layer containing a copy of the image with the blending mode set to one of the contrast modes (Soft Light works well). The effect is a dramatic increase in contrast, with color saturation boosted as well.

Adding a new composite layer can substantially increase the file size, plus if you change the layers below, those changes won't be reflected in the composite layer. A useful trick is to simply add a new adjustment layer (any kind) with the default settings, then set the blending mode for this new layer. This surprisingly accomplishes the same thing as copying the image into a new layer without increasing the file size substantially, and it will automatically update as the layers below are altered.

Figure 142 shows a Curves adjustment layer added to a narrowband color image of IC 1396. The original image is at top. The middle image shows the additional contrast and saturation gained by adding an adjustment layer with the blending mode set to Soft Light. On the bottom is the same image with some final stretching done with the curve in the adjustment layer.

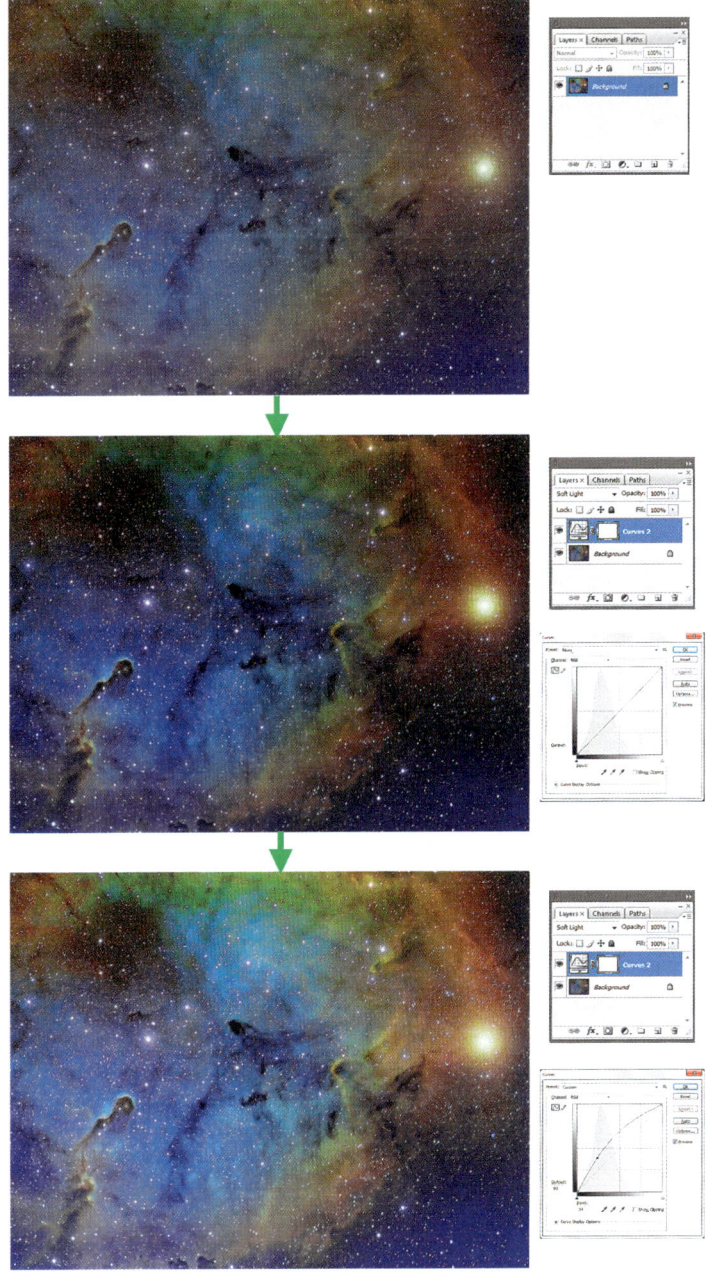

Figure 142. Using the Soft Light blending mode on an adjustment layer to boost color and contrast

Microcurves. As we've seen, the careful use of the Curves tool can also be used to create contrast within an object. Figure 143 shows how this can be applied in the final stages of processing to highlight differences not affected by other contrast enhancement techniques. The key is to set anchor points based on the specific levels in the image you want to target. Notice the improved visibility of the dim nebulosity on the left edge marked 'A' and the stronger contrast around the area marked 'B.'

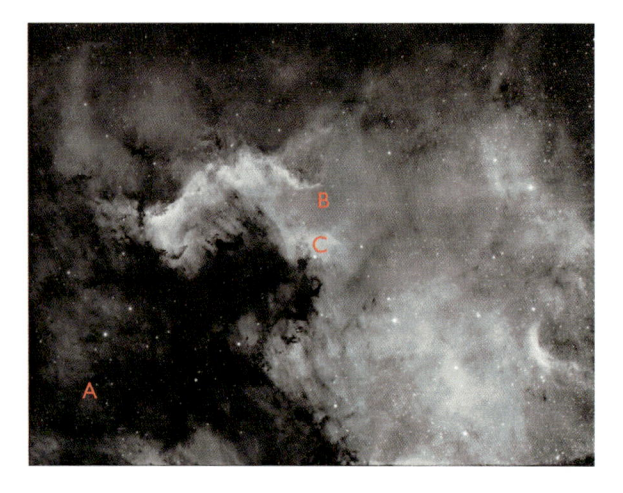

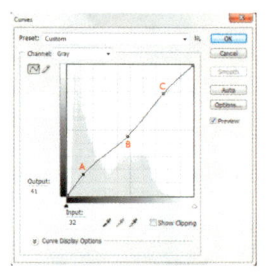

Figure 143. Using microcurves to create contrast between specific levels in the image

The High Dynamic Range Tool. Daylight photographers create high dynamic range images by merging multiple bracketed exposures, taking the shadow detail from the overexposed images and the highlight details from the underexposed images. The effect is to create a view with greater dynamic range than would have been possible to capture in a single exposure. With Photoshop CS5, Adobe introduced a tool called HDR Toning that is designed to create the look of a high dynamic range image without actually combining multiple images. The tool is a bit of a black box in terms of exactly how it works, but it combines curves, sharpening, and color saturation adjustments, with some filtering involved. Despite this, it can be used with astronomical images to bring out subtle details once the image processing is completed using standard tools.

If you have CS5 or later, the tool can be found under Image > Adjustments > HDR Toning. As with any drastic adjustment, copy the final image into a new layer before applying the tool so you can attenuate the effect later with layer opacity or masking. In the Edge Glow section, the radius setting works similarly to the high pass or unsharp mask filters. Try a large radius relative to the scale of the objects in your image, but keep the strength low. In the Tone and Detail section, you can generally leave the exposure, shadow, and highlight sliders alone. The gamma slider can help you apply the effects to the right levels in the image, just like with the levels tool. Be judicious with the details slider, as small changes here can drastically affect the image. Adjustments in the Color section are fine, but these are mostly redundant with other methods of color adjustment.

Processing Images

21 STAR ADJUSTMENTS

Star removal and star selection

Separating the stars to a separate layer or image can be very helpful, especially for images of diffuse nebulae. This allows stretching and other processing of the dimmer nebulosity alone that might adversely affect the stars if applied globally. It also allows you to adjust the stars separately, which usually involves shrinking them or increasing their color saturation. Separating stars from a more discrete object like a galaxy is fairly straightforward; however, when the stars are embedded in nebulosity, it can be far more challenging.

There are several approaches to this separation problem. Since the stars are identified in the registration and alignment stages of calibration, some programs will provide you with a star mask image that you can use for post-processing. This can be helpful, but sometimes the smallest stars are not visible before stretching, so the mask may contain only bright stars. Another approach is to use Photoshop's Dust and Scratches filter to remove the stars from a duplicate layer, then use other tools to select them, adjust them, and blend them back into the image. This is the most common approach, though there are many variations.

The Dust and Scratches method is approximately as follows. Each image requires its own settings. The figures show the results of a range of setting for one image.

1. Once enough stretching has been done to reveal all stars, duplicate the image to a new layer. This is a disposable layer that we will only use to remove the stars. A composite layer (Ctrl-Alt-Shift-E) works well for this if there are multiple layers already.

2. Select the Dust and Scratches filter (Filters > Noise > Dust and Scratches). Set the radius to the lowest possible value that removes the majority of stars. Settings that are too large will cause a loss of detail and posterization of levels.

   » In Figure 144 the original image of IC 2177 is at top, with radius settings of 10, 20, and 30 pixels shown below it. At 10 the stars are not fully removed, while at 20, they are, but at the expense of significant posterization (see the tool's magnified view). At 30 there is substantial loss of detail. After some experimentation, a value of 15 pixels was used as the best compromise.

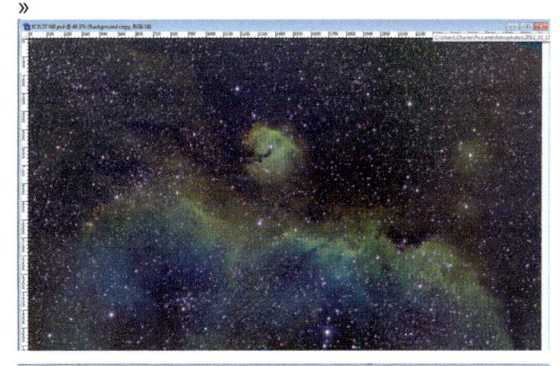

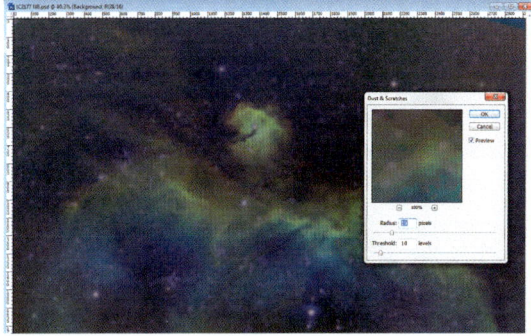

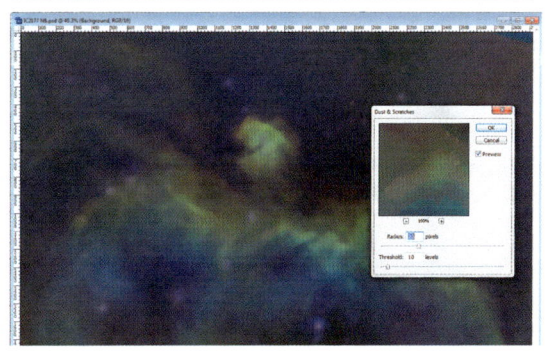

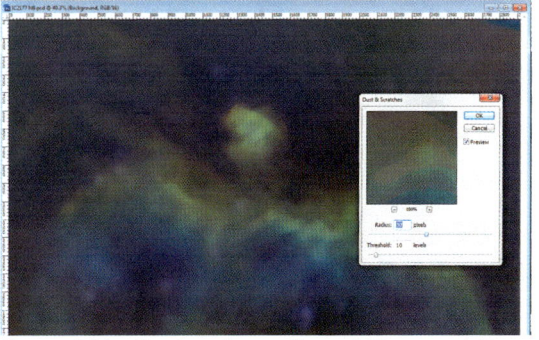

Figure 144. Setting the radius of Dust and Scratches tfor star removal

155

3. Set the threshold to the highest value that has minimal or no "rings" remaining where the stars were. Too low of a value will lead to posterization, and too high will leave the outer remnants of stars.

   » In Figure 145, again the original image is at top, with threshold settings of 21, 30, and 50 shown below it. At 21 there are no star artifacts, but there is some loss of detail. At 30 faint rings are starting to show, and at 50 they are prominent. A value of 28 was chosen as optimal here.

4. This starless layer can be used on its own for subsequent processing, or to retain more detail, change the layer blending mode of the starless layer to Darken. This will reveal the darker of the top layer or what is beneath it. This should erase the stars without affecting the rest of the image. If the filter has affected structures other than stars, you have to correct this manually or go back a step in the history palette and try different filter settings.

From here, you can proceed to process the starless image separately. To create an image with only the stars that can also be processed on its own, use the Difference blending mode. Set the starless version (you may have to create a composite layer for this) on top of the normal version of the image, with the layer blending mode set to Difference. The result should be black with the exception of the stars. If there are any non-stellar details, you can manually move them to the starless object layer or go back and adjust your star removal settings.

Figure 146 shows the stars-only result for the previous example. If only color adjustments were made to the stars, the two images can be brought back together with the Screen blending mode.

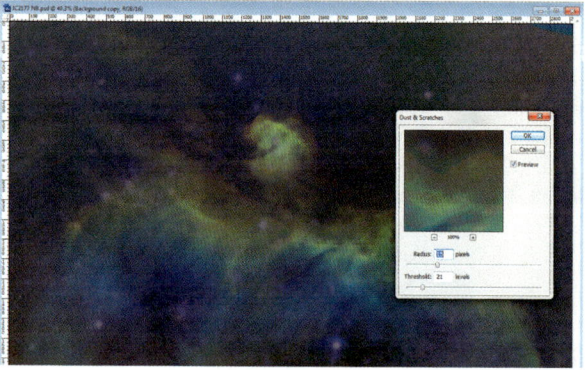

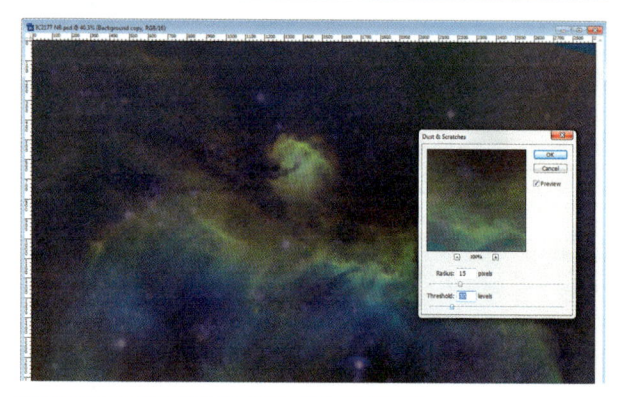

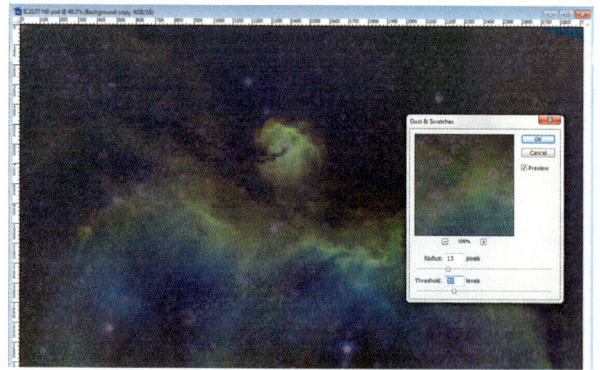

Figure 145. Setting the threshold of Dust and Scratches for star removal

Processing Images

Figure 146. Only the stars remain after subtraction with the Difference mode

If the stars were reduced or darkened, they will need to be selected to bring them back into the image, otherwise they will have halos. These can be pasted back onto the original image, or even onto a starless image. To work with the stars as a selection:

1. Create the stars-only image using the Difference blending mode shown above.
2. If there are faint non-stellar details, these can quickly be clipped out with the black point slider of the Levels tool. Since we are only interested in using this layer to define a selection, we can be destructive to any artifacts.
3. Desaturate the stars (Image > Adjustments > Desaturate). This will help in the next step. Remember, we aren't keeping any content from this layer; we're only using it to define a selection we can use on other layers.
4. Use the color range tool (Select > Color Range) to select of the stars. Zoom in and click on one star with the fuzziness set fairly low. Change to the additive ("+") dropper or hold Shift while you click to add new shades to the selection. This is shown in Figure 147. If you make a mistake in your selection, undo it with Ctrl-Z.

> If the "marching ants" selection outline interferes with your view, Ctrl-H will hide it.

5. Once all of the stars are selected, save the (Select > Save Selection) selection for good measure in case we don't like our results after the next step.

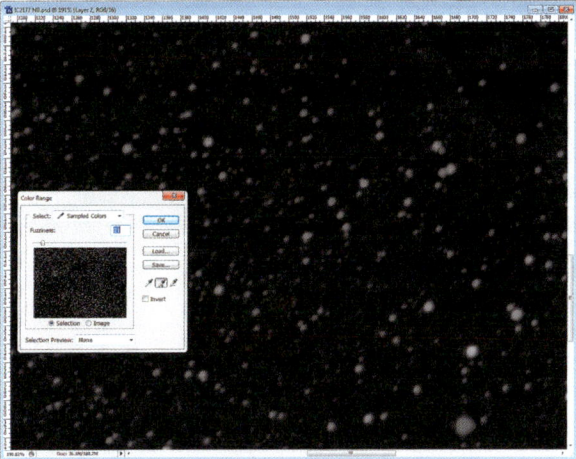

Figure 147. Using Color Range to select stars

6. Expand the selection a couple of pixels (Select > Modify > Expand) to ensure we aren't missing the rims of any stars, then feather the selection (Select > Modify > Feather) about half as much as the expansion.
7. Now, with the selection still active, click the original image layer. The stars will now be selected from there.
8. Copy the selection to its own layer (Layer > New > Layer via Copy). The result is shown in Figure 148.

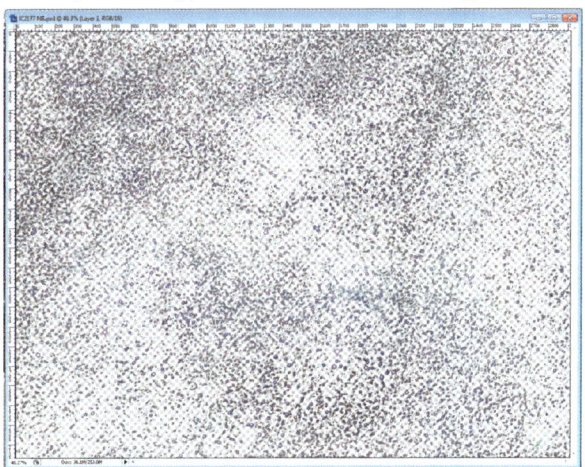

Figure 148. The stars selection on separate layer

9. This stars-only layer can now be processed as needed. Usually this is a boost to color saturation and reducing their size.
10. To bring it back to an image, no special layer blending modes are needed. The background is transparent, so simply put the stars layer on top. If this creates holes in the brighter regions of the image, you can mask those areas.

A quick-and-dirty method for star selection is to skip the creation of a starless image via Dust and Scratches and try to select the stars directly in the original image using the Color Range tool. This works better for some images than others. In many cases, selecting the object along with the stars in unavoidable, as they share similar brightness and color. If parts of your target object are also selected, choose the lasso tool and hold down on Alt while encircling any areas to be removed from the selection. Again, expand and feather the selection a few pixels.

Reducing star size across the image

Especially for wide field images of objects in the Milky Way, thousands of small stars in an image can overwhelm the deep-sky object, making it seem to fade into the background. We want to ensure the background stars are not the center of attention, but merely dimming them is not the answer. We need to shrink them. Not only does this put the emphasis on the object, but tight stars are also one of the visual cues for a "sharp" image, lending an impression of focus across the field of view. Even when our actual focus or tracking was a little off, we want smaller stars.

There are several methods to shrink many or all of the stars in an image at once, though in general it is best to do this in the final steps of image processing, or at least after all of the stretching has been completed. The Minimum filter (Filter > Other > Minimum) is one powerful tool for adjusting stars, though it has to be applied judiciously to avoid erasing stars.

Functionally, the Minimum filter is very simple: it replaces the value of each pixel with the dimmest one found within the selected radius. This increases the amount of dark area, shrinking lighter areas. For stars, it basically trims one pixel off the radius and dims the remaining pixels. With a one pixel radius setting (the smallest Photoshop allows), any star 2×2 pixels or smaller will disappear. A 3×3 star would be reduced to only the center pixel, and that would be reduced in brightness to the dimmest of the star's edge pixels. Since the effect is linear, it has a greater effect on small stars. A large 30-pixel diameter star would be reduced to 28 pixels across, but a four-pixel diameter star would become two pixels wide.

Figure 149 shows the effect of the Minimum filter on a 1000×1000 pixel area with settings of one, two, and three pixels. Even at this scale, settings of two and three pixels are erasing stars and the existing posterization is emphasized. The filter's effect can be faded via layer opacity, but it's still best to keep the radius setting as small as possible.

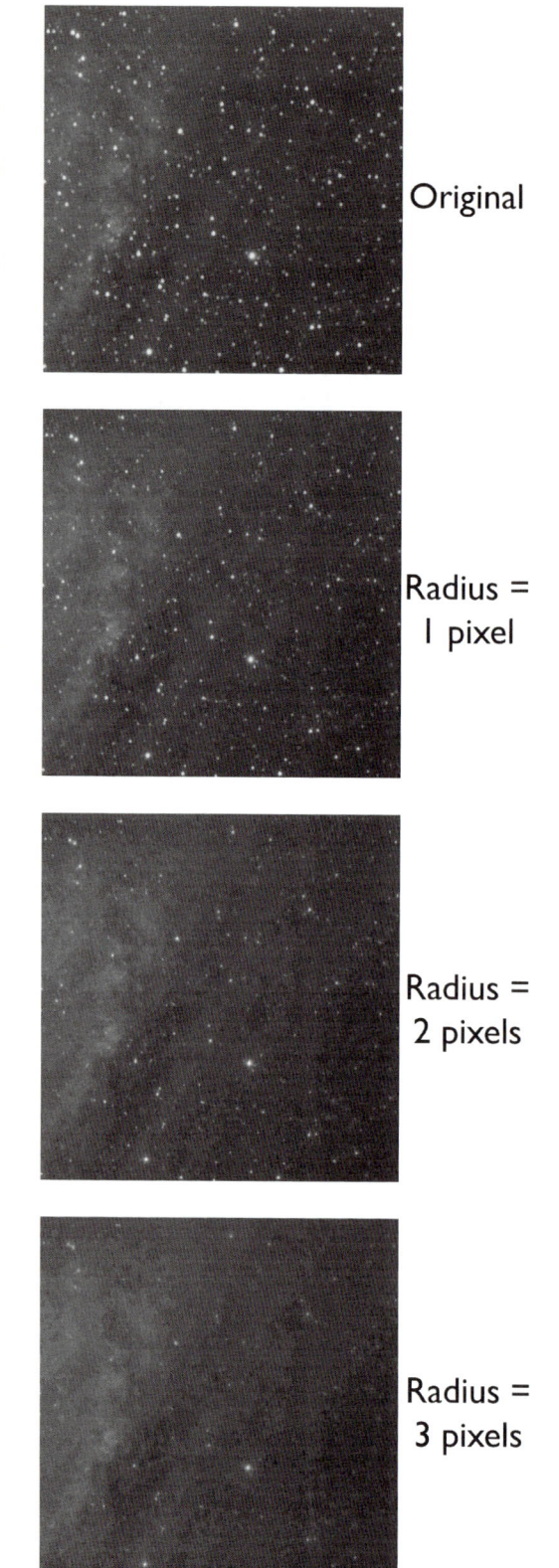

Figure 149. The Minimum filter

Processing Images

Whenever possible, we don't want to apply the Minimum filter to the whole image, as that chips away at the edges and brightness of nebulosity. (Since they have no nebulosity, globular clusters are one exception.) If you already have a star mask or the stars are already isolated to their own layer, apply the filter there. Any star selection or masking technique can be used, but it is worth the effort to separate the stars before applying the Minimum filter. This avoids damage or artifacts in the rest of the image.

Figure 150 shows a detail of an image where the Minimum filter was applied the whole image. Notice the unnatural star shapes that are created within the nebulosity. Applying the filter only to the stars would have avoided these artifacts.

Figure 150. Star-shape distortion from the Minimum filter

Even at its smallest setting of one pixel, the Minimum filter is too strong for images shot at short focal lengths or with large-photosite sensors. If this is the case, there is a way to apply it with a fractional radius—increase the size of the image before applying it. Since data are lost when pixels are fractionally resampled, scaling up by an integer amount is preferred. Doubling the image size, applying the filter with a one pixel radius, then reducing the image size back in half will result in the equivalent of a 0.5 radius application of the filter.

For many images, you will want to select all of the stars for reduction, as this can do a lot visually to bring the object of interest to the foreground. Other times, only a few of the brightest stars need to be put in their place. If the stars you want to shrink are embedded in nebulosity, more control is usually needed, so you may have to use the oval selection tool (feathered, of course) to manually select them. This is covered in the next section.

Reducing the brightest stars

The brightest stars in an image can become bloated in long exposures, and sometimes no amount of control in the stretching process can prevent it. One of the best tools to reduce the size of individual stars, and even reflection halos, is the Spherize filter (Filter > Distort > Spherize). Select the star and its halo using the elliptical marquee tool (hold down shift while dragging to select a circular area), then feather the selection. Ensure that the selection is centered on the star and that there are no other stars within the selection, as any off-center stars or features will be stretched toward the center. If selecting other stars is unavoidable, you may have to erase them or copy them back in after applying the filter. Once you have the selection, open the Spherize filter and slide the amount slider left to a negative value that shrinks the star the desired amount.

Figure 151 shows a small region around a problem star that is not only too bright, but suffers from a substantial reflection halo. The selection around the star's halo has been feathered, but it is impossible to avoid two smaller nearby stars. The Spherize tool's preview shows how those two stars will be stretched.

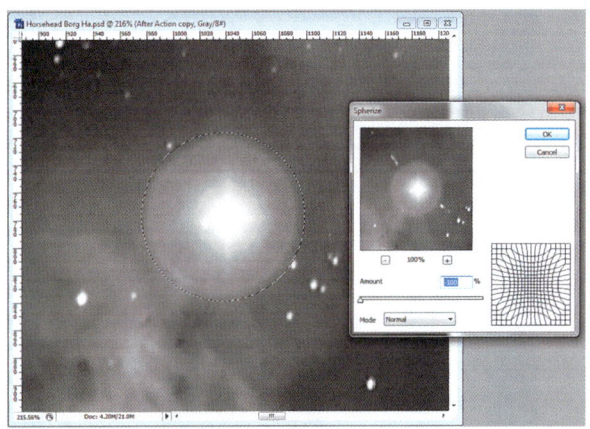

Figure 151. The Spherize filter

Figure 152 shows the result after two consecutive applications of the tool, followed by restoration of the two small stars by pasting back in them from the pre-distorted image. If the halo is still too prominent, a quick adjustment was Curves while the selection is still active will darken it.

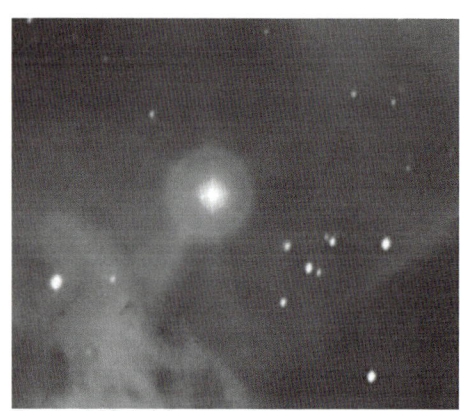

Figure 152.   After the filter

Unfortunately, Spherize and many other tools in the Distort filter group were 8-bit only until CS5. This limits their use in CS3 and CS4 to final adjustments, and only then if an 8-bit output is acceptable.

A newer tool for individual star reduction is the Liquify tool. In Photoshop CS5 and later, the tool has a pucker setting that can be used to quickly shrink stars with a brush tool that eliminates the need for creating selections. The time the mouse button is held down determines the strength of the effect, making it an effective tool to quickly shrink many stars. The brush rate setting controls the level of effect applied per second.

Correcting star color in narrowband images

A common problem in many mapped-color narrowband images is magenta stars or magenta star halos. This is a result of the greater stretching usually applied to the OIII and SII channels in order to reveal their dim nebulosity. This stretching affects the stars as well, so the stars in these channels end up brighter than those in the H-alpha channel, which does not need to be stretched as drastically for most objects. When using the Hubble palette, H-alpha is assigned to green, so after color synthesis, the stars end up with a dearth of green, resulting in a magenta or red hue. The brightness of stars in the H-alpha channel can be directly suppressed via the Minimum filter or other methods, but the stars are usually embedded in nebulosity, making it difficult to isolate them.

If there are no other areas in the image that are magenta, the selective color tool (Image > Adjustments > Selective Color) can be used to remove the magenta. Choose Magentas from the drop down, and add green back to restore proper color. If the selective color tool doesn't seem to target the right color, you can use the Color Range tool (Select > Color Range) to directly select the bothersome color. Expand and feather the resulting selection just a couple of pixels, then use the Hue/Saturation tool to correct the hue, or simply desaturate it.

In many cases the star halos have the same color as desirable nebulosity elsewhere in the image, making their selection nearly impossible. For small stars there is an easy solution to color halos: they can be removed with the Reduce Noise filter's "Reduce Color Noise" slider. Figure 153 shows a case of red star halos that stand out in an area of green in a narrowband mapped-color image.

Figure 153.   Magenta star fringing in a mapped color image

Figure 154 shows the result of Reduce Noise with the Reduce Color Noise setting at 70%. The other sliders can be set to zero if luminance noise reduction is not needed. Using selective color methods would have desaturated reddish regions of nebulosity elsewhere in the image.

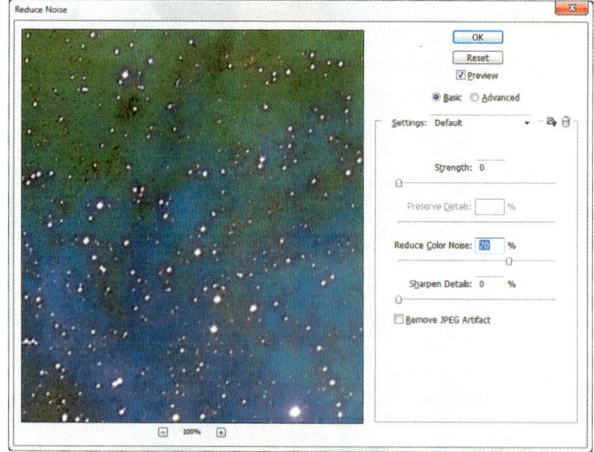

Figure 154.   The Reduce Color Noise setting can completely remove color fringing

Any of the above methods creates neutral star colors, not the colorful stars seen in RGB color images. The only way to ensure there are proper star colors in a mapped-color narrowband image is to copy them from an RGB image. While this adds a significant processing step, the additional exposure time needed to capture star color is minimal. Since you are only exposing for stars, and RGB filters pass far more light than narrowband filters, a few short exposures for each channel is enough to create an image with star colors. After selecting the stars in this image, copy them into a new layer in the narrowband image with the blending mode set to Color.

22 NOISE REDUCTION

Visual noise, color, and scale

Noise visually manifests itself in two ways: luminance noise and chrominance noise. Both are variations in brightness, but the visual effect is different because chrominance noise acts in the color channels, so it appears as random variation in color, usually speckling or blotching. Chrominance noise can be an artifact of the de-Bayering process, but color images synthesized from filtered exposures also suffer from it. Luminance noise is variation in the brightness of pixels, even if we view them in monochrome. Figure 155 shows an example of chrominance noise.

As we've discussed, color changes on a larger spatial scale than luminance. Chrominance noise stands out because it is variation in color at the smallest scale. Fortunately, this difference in scale between color and luminance means we can apply aggressive correction to chrominance noise without affecting the details in our image. Luminance noise, however, is more difficult to tackle without sacrificing detail. It represents the underlying signal-to-noise ratio, which can only be improved by increased exposure time.

Figure 155. Chrominance noise in the background of an image

Noise is also related to image scale. We've seen that binning our exposures improves the SNR by gathering more light per pixel at the expense of having larger pixels, and thus less resolution. This relationship between image scale and noise is also true (though for subtly different reasons) once the image is processed. Blurring or reducing an image's resolution effectively averages the values of across some range of pixels. Visually, this creates smoother gradients and blurrier details, but it also generates a more accurate estimate of the true brightness for a given area by incorporating more data into the average. This sacrifice of detail improves the SNR. This is one of the reasons why a mediocre image may look great as a thumbnail, but looks progressively worse as you enlarge it.

Tools for noise reduction

Most image processing packages offer dedicated noise reduction tools, each with its own algorithms. All of them work on the principle of finding outlier pixels and adjusting them so they are more similar to their neighbors. This smoothing can cause blurring, so the goal is to strike a balance between retaining fine details and reducing noise. Under the hood, the mathematical modeling used can be very sophisticated, with some programs taking advantage of the high spatial frequency of image noise by using wavelets to attack the problem. Noise reduction can do wonders for an image, but there can be a small learning curve with each tool to find the parameters that work best.

While Photoshop offers its own proprietary tools for noise reduction, there are many commercial programs and Photoshop plug-ins available. These include Noise Ninja, Neat Image, DeNoise, and many others. Dedicated astroimaging applications also offer their own noise reduction tools. PixInsight offers the excellent ACDNR (Adaptive Contrast-Driven Noise Reduction). IRIS has a wavelet-based noise reduction tool Nebulosity implements the GREYCstoration algorithm among others. Both MaximDL and ImagesPlus offer proprietary smoothing and noise reduction tools.

Two other approaches to noise that work especially well when we can isolate the background are to desaturate the background, which eliminates chrominance noise, or to blur the background slightly. Neither is appropriate for application to an entire image. Figure 156 shows a patch of background with both of these approaches applied, as well as two of Photoshop's noise reduction tools: Reduce Color Noise (part of Filters > Noise > Reduce Noise) and the Despeckle filter (Filters > Noise > Despeckle).

Bear in mind that you don't want a perfectly black, noise-free background. Small scale noise will disappear when the image is viewed at a normal magnification, and it doesn't look realistic to have a background that is too dark or too smooth. Noise is variation above and below a mean, in order to show variation on both sides of this mean, the background brightness has to be above zero. Neutral RGB values in the range of (10, 10, 10) to (40, 40, 40) look natural

Processing Images

in most images. If the black point is pushed too high, the effect on the background is to create black patches wherever the noise was clipped. Figure 157 shows this. Always maintain the left "toe" of the histogram for a realistic-looking sky background.

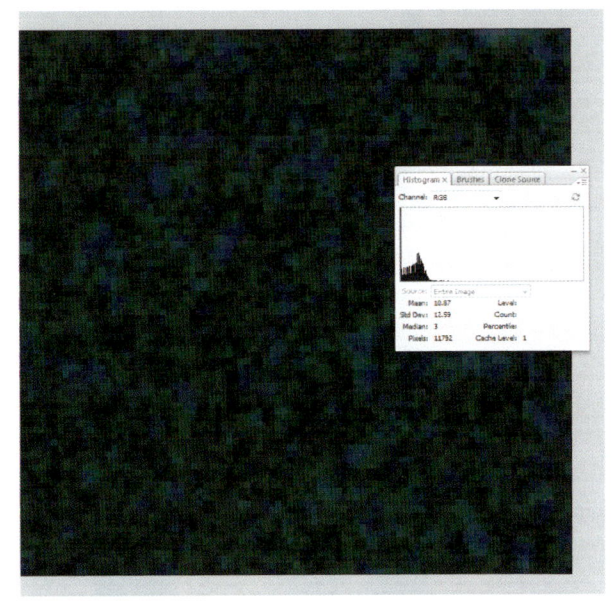

Figure 157. Black point clipping of background noise

Applying noise reduction selectively

We use noise reduction at the other end of the signal-to-noise ratio spectrum from sharpening. Noise reduction is needed most in the dim regions where the SNR is lowest. If noise reduction is applied to the whole image, important structural details can be blurred. As with sharpening, we need to selectively apply our noise reduction, but to the dimmest areas instead of the brightest. To do this, we follow the same procedure we used for sharpening, but we invert the layer mask. The goal is to create a top layer with noise reduction applied that will sit over the existing image. We can then selectively reveal more or less of each layer based on their brightness at each pixel. As with sharpening, noise reduction requires the use of a filter, so it cannot be applied as an adjustment layer. The image needs to be nearly final, because we are going to act directly on it (or a copy of it).

11. Duplicate the image into a composite layer (Shift-Ctrl-Alt-E).
12. Apply the noise reduction filter of your choice to this new layer.
13. Create a layer mask on this new top layer by clicking the layer mask icon on the Layers palette. Copy the image to the clipboard (Ctrl-A, then Ctrl-C).
14. View the mask by Alt-clicking on the thumbnail, then paste the image onto it (Ctrl-V). Since black reveals what's below in a layer mask and white shows what is in the top layer, we need to invert our image to reveal the top (noise-reduced) layer where the pixels are darkest, and keep the bottom layer where they are is

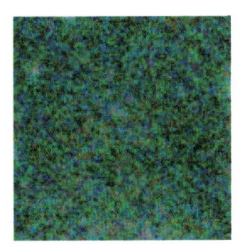

Original

Gaussian blur 0.5 pixels

Desaturate 50%

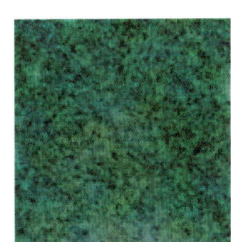

Reduce Color Noise 50%

Despeckle

Figure 156. Four methods of background noise reduction

brightest. With the mask visible, invert it with either Ctrl-I or Image > Adjustments > Invert.

15. Shift the balance of the noise-reduced top layer and the existing image by adjusting the layer mask via a Levels adjustment.

16. If there are particular areas that need the most noise reduction, you can paint white directly onto the layer mask.

If you only want to desaturate the background, the procedure is simpler:

1. Copy the image to the clipboard (Ctrl-A, then Ctrl-C).

2. Create a new Hue/Saturation adjustment layer.

3. Add a layer mask to it by clicking the layer mask icon on the Layers palette.

4. View the mask by Alt-clicking on the thumbnail, then paste the image onto it (Ctrl-V).

5. Shift the balance of the noise-reduced top layer and the existing image by adjusting the layer mask via a Levels adjustment.

6. Mask off any areas that are not to be desaturated at all by painting black directly onto the layer mask.

Processing Images

23 COMPOSITION

Framing the scene

The object of interest does not need to be centered, nor do you need to show the whole object. Consider it the same way you would a portrait of a person. The best portraits capture a subject's character by framing them in unique ways or showing them in a particular context. A portrait with the face centered squarely is simply a mug shot. The "rule of thirds" used for centuries by painters and photographers also applies to astroimaging. Points of interest in an image are better placed where one-third lines intersect than directly aligned with the horizontal or vertical middle of the image. In the same vein, try to keep stars or points of interest off the very edge of the frame. This can be distracting, as it looks like the image was haphazardly cropped.

Also think about scale. Rarely do you want a small, lone galaxy in a sea of black. The result can feel clinical or expected, like an image in a textbook. The most impactful images waste no space. Fill the frame with the object of interest if you can, even showing only part of it. As an example, consider the IC 1396 nebula complex. While the whole complex is stunning, sometimes showing only part of it makes the most sense, as in Figure 158, where the "elephant trunk" feature and Mu Cephei (the bright yellow star at right) can stand out as compositional elements.

Figure 158. Sometimes the best framing of an object is to show only part of it

When choosing the final scale and orientation of the image, bear in mind that resizing or rotation transformations involve resampling, which can lose image data, especially if applied repeatedly. Make these changes at the final stage of processing and only if necessary. Obviously, downsizing the resolution of an image, usually for web presentation, is a destructive process, but upsizing can also require resampling that can lead to small losses in fidelity. Image rotation to an angle other than 90, 180, or 270 always requires resampling.

Sometimes, when nothing else works, break the rules and celebrate symmetry. Figure 159 shows the Rosette Nebula. The rule of thirds doesn't work here, because there are no nearly objects in the field to balance the image and the nebular has a very nearly circular shape. In this case the Rosette's three-dimensional structure is enough to stand on its own.

Figure 159. Then again, sometimes symmetrical framing is okay

Orientation

When considering orientation, it is mostly a matter of taste, but our minds are used to inferring motion and perspective, even where there is none. Deep-sky objects are at such great distance that there is no true perspective or infinity point to create a sense of depth, so these inferred bits of context are important. Figure 160 shows four orientations of the same image of NGC 1499. The top two seem to imply (to the author at least) that the nebulosity is being lifted from beneath (upper-left image) or from the left (upper-right image). Ultimately, the upper-left orientation was chosen for the cover of this book, as it looked the most "stable" from a purely subjective perspective. Whether or not you agree with that choice, it should be clear that each orientation conveys a very different feel.

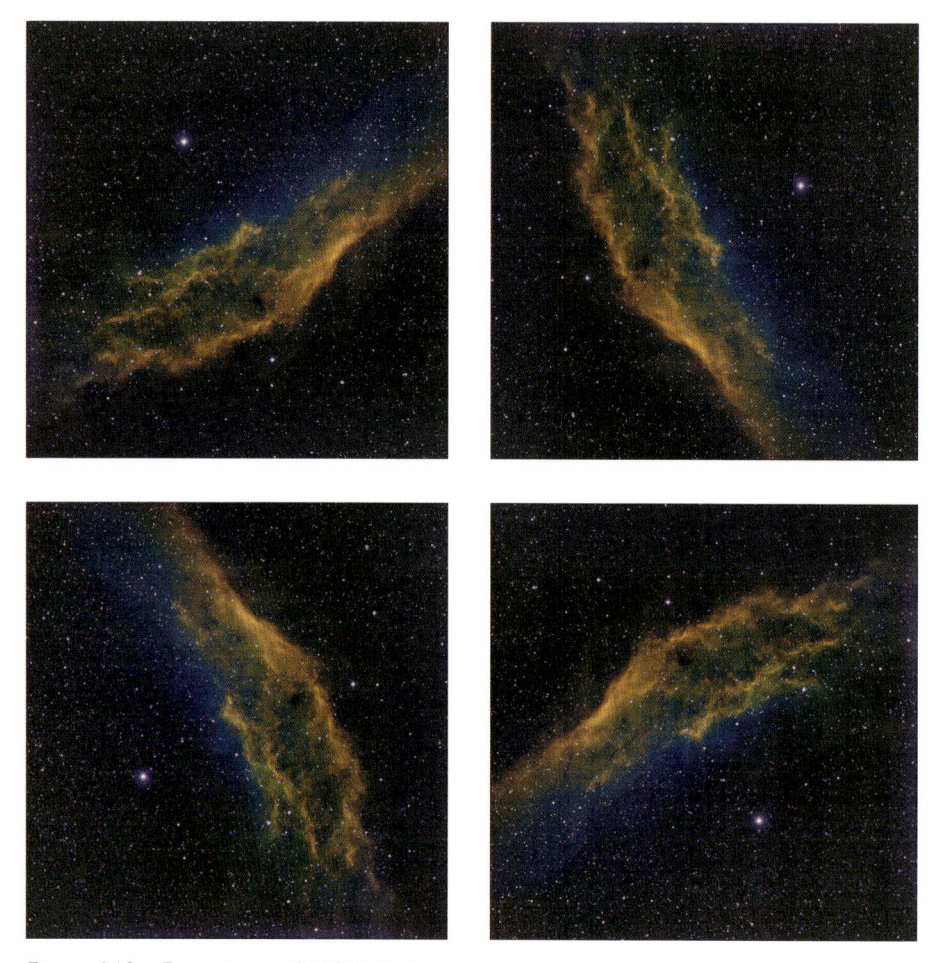

Figure 160. Four views of NGC 1499

Spiral galaxies are somewhat unique in that they have a known three-dimensional shape. Perfectly edge-on or face-on galaxies can be presented in any way, but those viewed from an intermediate angle have a clear front and back. In this case, the near side of a galaxy should be oriented to the bottom so the viewer has the perception of looking at it from slightly above, as this is the expected perspective for most objects we see. If the image is shown with the near side toward the top, it presents the galaxy from the odd perspective of looking "up" at it from underneath.

Differences in sharpness are a strong visual cue for distance. This effect of our atmosphere, that objects further away fade into the haze, is incorporated into the brain's visual heuristics. We can take advantage of this to make a galaxy image appear more three-dimensional. The front of a galaxy should generally be sharper than the distant side, as there is less intervening dust to obscure our view of the front. Emphasizing this by applying a little more sharpening to the front side can enhance the sense of depth.

Color

People respond to color. Highly saturated images with strong color contrast attract more attention and are generally more engaging. While some deep-sky objects look beautiful in monochrome (especially H-alpha) because of their texture, even those images frequently look better in color. There are exceptions to every rule, but if you're not convinced that this is the case, look through the galleries in major astronomy publications or on NASA's Astronomy Picture of the Day website and ask yourself how many of those images are black and white.

Walk away

Periodically, walk away from the computer. Take a break from processing. Come back after a few minutes. This has nothing to do with being tired, though fatigue doesn't help.

Figure 161. Which version of M63 is right-side up?

It's important because your eyes become acclimated to the image, the composition, and especially the colors. After a while, you are no longer seeing it as it really is.

Remember that colors are created in the brain. What you swear is cyan may reveal itself as more of a blue when you come back. The image may look duller than it did five minutes ago. You may realize that the last application of sharpening was a step too far. Walk away for a minute and come back with fresh eyes. Great visual artists have done it for centuries with good reason.

24 DSLR Processing Example: The Witch's Broom Nebula

Now, let's put processing our tools together with a few start-to-finish examples. The Veil Nebula complex in Cygnus is notoriously difficult to see visually, but it offers beautiful photographic opportunities to imagers in the northern hemisphere every summer. Because it is an emission nebula with most of its light at the H-alpha wavelength, DSLR users will achieve much better results with a camera modified to replace the IR filter, as was the case with this image.

The complex is an ancient supernova remnant that covers hundreds of square degrees, but its fragmented nature makes it amenable to both wide-field and longer focal length views. Here, we will process an image of NGC 6960, the Witch's Broom Nebula. This is a small but photogenic fragment of the complex that passes through the magnitude 4.2 star 52 Cygni. It emits light primarily at the H-alpha and OIII wavelengths, creating contrasting blue and red tendrils in images.

Relatively modest equipment was used to capture the image. Six hours of exposures were taken through a 120 mm f/7.5 doublet refractor (Orion's EON 120) using a Canon Rebel XSi/450D DSLR with the stock IR filter replaced by a clear filter. Each exposure was four minutes long at a gain setting of ISO 1600. Autoguiding was done with a Meade DSI fitted to a 50 mm finder scope.

Because of the refractor's sensitivity to temperature, focus degraded substantially as the night cooled, so only three hours of data (45 four-minute exposures) were calibrated. This is a very short exposure time, especially at f/7.5, and additional data would reveal deeper details in the image. We'll see the limits of these data as we process. But the nebula was nearly at the zenith during these three hours and the transparency was excellent, so three hours provides a good result.

The 45 light frames were calibrated with DeepSkyStacker using a master dark frame composed of 60 exposures at nearly the same temperature from a library of darks taken earlier in the summer. About 50 flat frames were used, and the master bias frame was also from a library of earlier exposures, composed of over 100 bias exposures. The exposures were stacked using the median kappa-sigma method (kappa = 2). The "RGB Channels Background Calibration" option was selected, so the color channels were nearly balanced in the calibrated image.

Right out of calibration, the image shows a lot of potential. The nebula is clearly visible even before any stretching, which is a good sign. Initially, there is no color saturation and the contrast is washed out, but we know we can fix that. 52 Cygni is very bright, with a halo around it, so we'll have to keep an eye on it as we stretch the image.

Processing Images

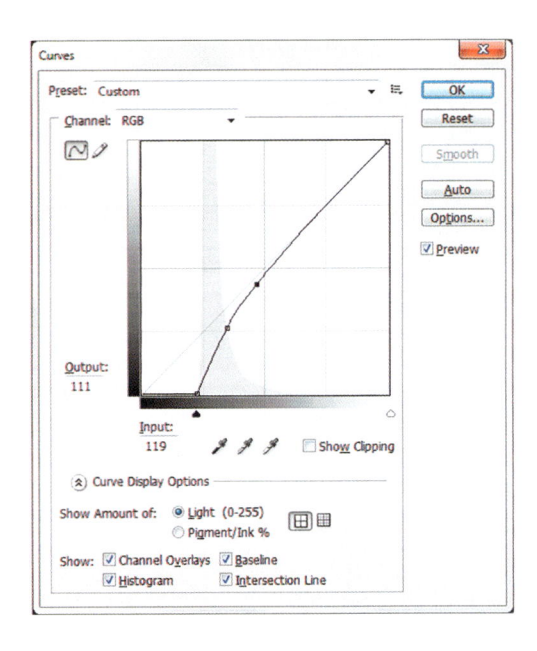

A quick look at the histogram shows that a quarter of the dynamic range is being wasted, so the black point needs to be moved up. The first curve applied is mainly to accomplish this, along with some gentle stretching to the midtones. Recall that the line-of-no-change runs from the black point to the white point, so in this case, the 45° line provided by Photoshop is not the reference line.

The lower anchor point on the curve is based on a sample of the nebula's brightness. (Hold Ctrl while clicking on the image to set an anchor point.) The higher anchor point is there to flatten the curve for the brighter levels, keeping the stars from getting too bright.

After this first stretch, things are already looking better. The background is darker and the nebula is brighter.

On the left side, the image is lighter— a calibration artifact where exposures must have overlapped. Fortunately, this isn't on part of the image we are interested in, so we can crop it out now (step not shown). If it were in a critical part of the image, we'd have to go back to our raw images and find where this overlap problem was coming from.

Since we know that stretching desaturates images, let's boost the color saturation before we apply any additional curves.

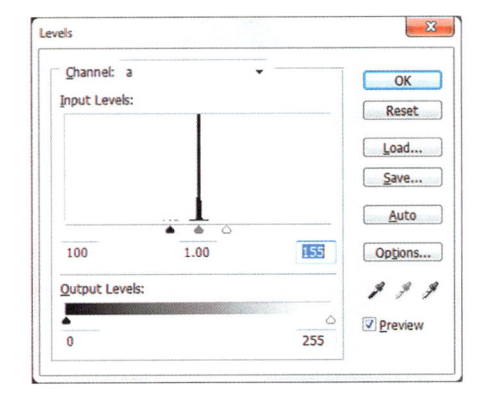

After changing the image mode to Lab Color, we can use the Levels tool to dramatically expand the dynamic range of the a and b channels. It is important to set the black and white points (though these terms are not really accurate when speaking about Lab's two color channels) symmetrically on each side, and equally to both the a and b channel.

Once the saturation boost is done, it's safe to change back to RBG mode.

169

The colors are dramatically improved. The red and blue areas of the nebula stand out, and the background stars are full of color. Along with this however, comes some color in the halo around 52 Cygni and greater chrominance noise in the background. We'll come back and fix these later.

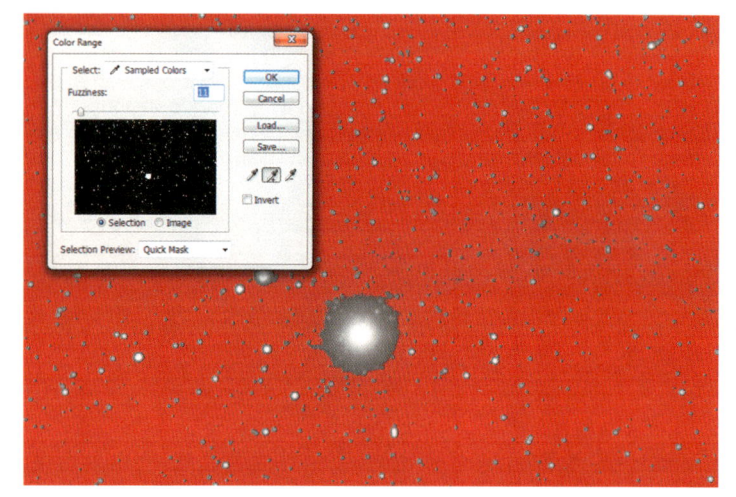

While the star colors are great, we want the nebula to be the center of attention. We need to reduce the prominence of the stars. To do this, first, create two copies of the image in new layers, then desaturate one of them. We'll use the Color Range method of star selection, so desaturating a copied version of the image makes this easier.

Now, use the color range tool's additive dropper to select a few stars. With the Quick Mask active on the drop down menu, zoom in on part of the image with nebulosity, and then slide the Fuzziness slider until most of the stars are still selected, but the areas with nebulosity remain masked red as not part of the selection.

If any areas of nebulosity slipped in, we can use the lasso tool to deselect them by holding Alt while encircling them. Save the selection for good measure, then expand it by one pixel and feather it by one pixel. With the selection still active, select the copied layer that you didn't desaturate and apply the Minimum filter to the selection. The result is shown at left. The stars are much less prominent, but the nebulosity is unaffected. The desaturated layer can now be deleted, and you can see the difference by toggling the new layer. To fade the effect, adjust the opacity of the new layer, but note that several subsequent steps will brighten the stars again.

Processing Images

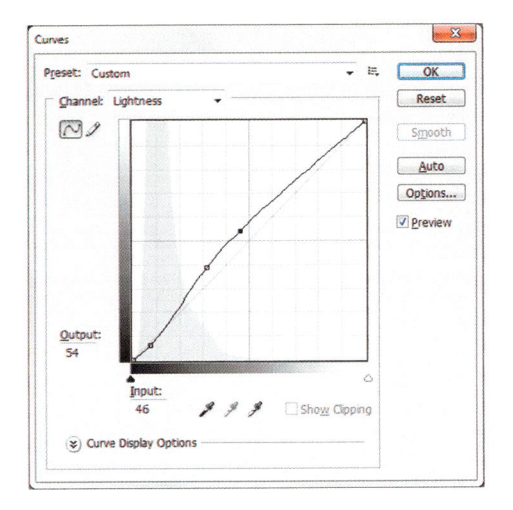

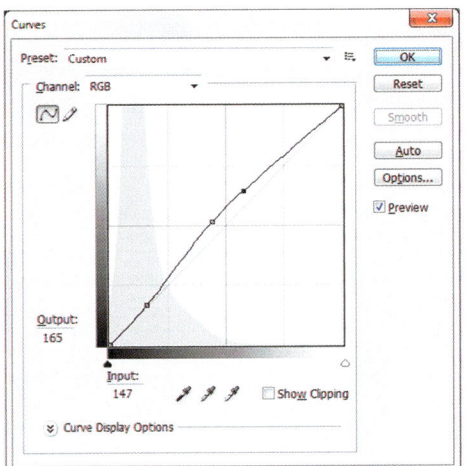

The image is still a little flat, so now that the colors are improved and the stars are reduced, we can apply another curve or two (via adjustment layers if you'd like) to take some of that dynamic range away from the highlights and give it to the nebula's midtones.

The two curves we use here are applied in sequence, and they are similar to the first curve, except this time we use three anchor points. The middle anchor is based on the nebulosity, and it is raised to increase its brightness. The lower anchor holds down the background, creating a subtle S-shape to increase contrast between the nebula and background. The higher anchor is again there to keep the highlights from getting too bright, because the more we push down on that lower anchor, the more the curve wants to bow out on the top.

We need to tame 52 Cygni a little bit. Rather than applying an inverted curve to dim it, we can simply take advantage of the fact that we still have the old layer with 52 Cygni before it was stretched. Draw a circular selection around the halo, about 200 pixels across in this case. (Remember that holding Shift while dragging will turn the elliptical marquee tool into a circular one.) Make sure it's centered over the star, then feather the selection 30 pixels.

Hold down on Alt while you click the new layer mask icon (or choose the Layers > Layer Mask > Hide Selection option). This will reveal the lower layer version of 52 Cygni.

The post-stretch top layer star is shown on the left, with the feathered selection in the middle (painted in black on the mask), which reveals the star on the pre-stretch bottom layer on the right.

Applying unsharp mask (after flattening the image) will highlight the nebula's structure and create contrast. Experimenting with the radius setting shows that a setting around 40 pixels seems to be the best match for the nebula. Unfortunately, it also greatly exaggerates all of the stars. Since the nebula has a clearly defined structure, the simplest way to isolate the filter's effect is to use another layer mask.

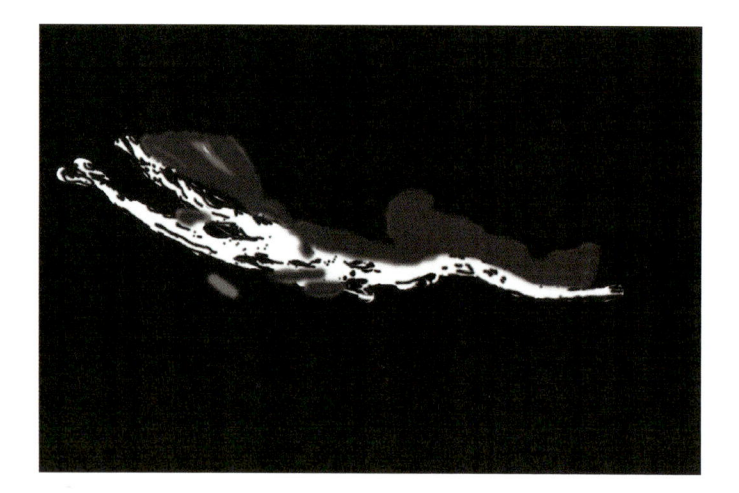

The mask isn't pretty on its own, but when used over the new unsharp layer, it allows us to "paint in" contrast to the features of interest.

Processing Images

The masked unsharp layer reveals the contrast-boosted nebula on top of the stars from the previous step.

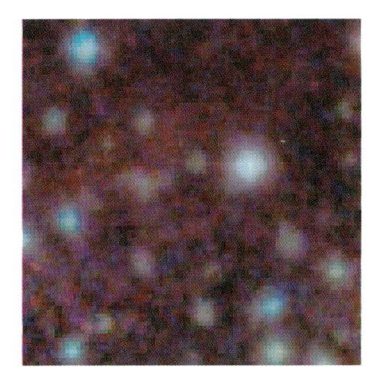

A closer look at the nebulosity reveals the limitations of our image capture (only three hours of exposures at f/7.5 with an uncooled camera in the summer). There is a good bit of noise at the pixel level. The chrominance noise is more visible and easier to fix without blurring the details of the nebular structure. Experimenting with the Reduce Noise filter shows that even moderate levels of noise reduction make the nebula look fake or plasticky, so a setting of 2 was used for overall noise reduction, with the Reduce Color Noise set to 45%. The results of this filter on a very small area are shown on the right.

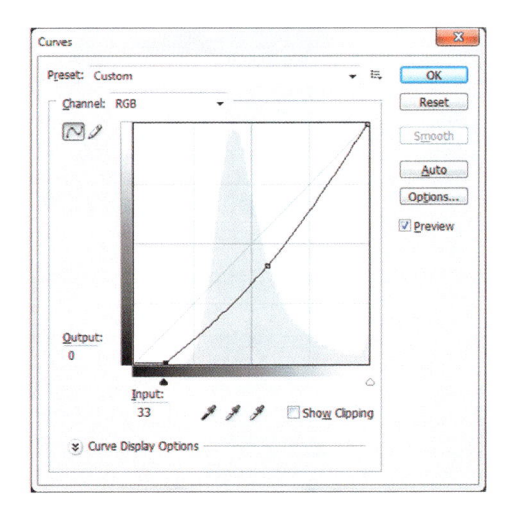

The stars could use a further step to push them into the background, especially given that many of the brighter ones have some haloing. This is a simple adjustment since we already have a saved selection that includes only the stars.

After loading the selection, it was expanded by one pixel and feathered by three pixels. The curve shown here was then applied to dim the selected stars slightly and differentiate their brightnesses by expanding their dynamic range via a black point adjustment.

173

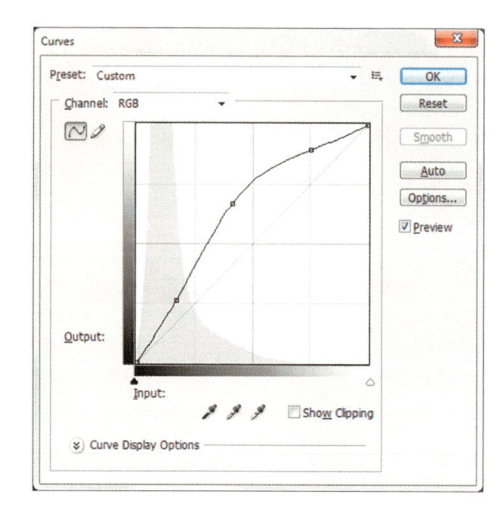

Finally, the image could use an overall contrast boost. We can do this via an adjustment layer set to one of the contrast blending modes. Recall that this saves us from copying the whole image to a new layer and pasting it over itself. We use a curves adjustment layer set to the Soft Light blending mode with the opacity set to only 30%. Further adjustments can then be made via the shape of the curve shown.

The final result is shown below and on the cover of this book.

25 CCD Processing example: The Rosette Nebula in narrowband

In wide field images of the Orion region, the Rosette Nebula stands out as a bright circular area west of Betelgeuse that is about twice the diameter of the full moon. Among emission nebulae, it is one of the brightest, with a complex and beautiful structure to match, making it one of the most imaged objects in the northern hemisphere sky.

To capture the full extent of the nebula requires a short focal length and a relatively large sensor. These raw subexposures were captured over three cold February nights with an SBIG ST-8300M through a 330 mm telescope (a Borg 77EDII) at f/4.3. In order to capture the faint narrowband signals, pixels were binned 2×2. The subexposures were 10 minutes each, with the sensor cooled to -20° C.

Twenty-one good H-alpha filtered exposures were taken, as were 19 SII and 15 OIII. Ideally, we'd have twice as much data for the dimmer OIII and SII channels as we have for H-alpha, but in this case a weather system brought weeks of clouds before another night of OIII or SII could be added.

The images were aligned on the best H-alpha subexposure. DeepSkyStacker's Auto Adaptive Weighted Average stacking method was used for the light frames. 20 flats were combined for each channel using the Median Kappa Sigma method. Because the images were substantially undersampled, the Drizzle feature was enabled at a 2× setting to help recover some of the lost resolution. Darks and bias frames were from a library, and over 40 of each were combined, also using the Median Kappa Sigma method.

This example maps the narrowband data to separate channels to create a false-color image. Obviously, the result is not what one would see through a telescope, but it allows us to clearly visualize the different emission lines in the nebula's light. From a processing perspective, nearly the same process would be used to combine red, green, and blue filtered images, excluding the Selective Color adjustments.

First, let's have a look at our calibrated and stacked H-alpha image. Before any processing has been applied, it looks pretty good. The Rosette's structure is visible, though it's obvious that some stretching is still needed. In the lower left corner, you can see a faint artifact where the subexposures overlapped (or didn't), but we'll crop that out after we bring the color channels together.

The SII image shows a substantial gradient from lower left to upper right, and the nebulosity is far dimmer, which we would expect. Overall the image is sharp.

The OIII image barely reveals any nebulosity, which is again expected before we've done any stretching. Unfortunately, the focus was also less than perfect here.

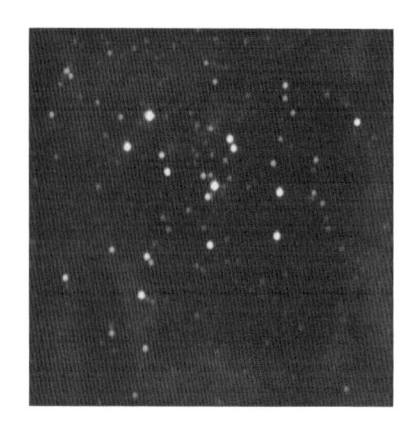

A close-up look at the calibrated OIII image (left) shows that the focus was not as good as the H-alpha image (right) or the SII image (not shown). If we don't correct the stars so they are approximately the same size across all channels, not only will the stars be too large in the final image, they will have noticeable color halos.

Processing Images

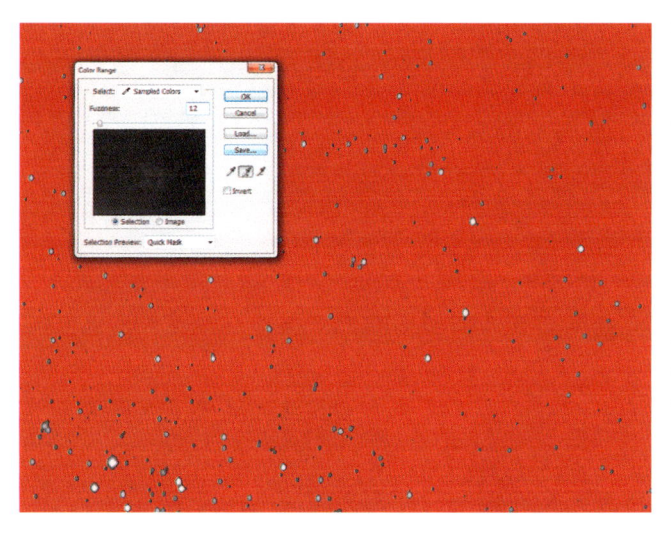

To bring the OIII stars into line with the other two channels, use the Color Range tool's additive dropper to select progressively dimmer stars down to the point where the stars are just brighter than the nebulosity. If the nebulosity becomes part of the selection, either undo the most recent additive selection, or dial back the fuzziness setting until there are only stars revealed on the mask.

Save the star selection, then modify it by expanding it one pixel and feathering it two pixels. If these values don't work, you can reload the selection and try new ones.

Finally, apply the Minimum filter to the stars with a radius of one pixel.

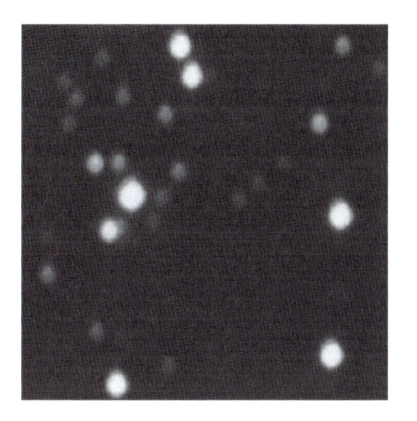

The left side shows a detail from the original OIII image, while the right shows the result after star minimization. While we didn't actually improve the focus, we have at least prevented the OIII channel from causing colored halos around the stars, and we did it without affecting the structure of the nebula.

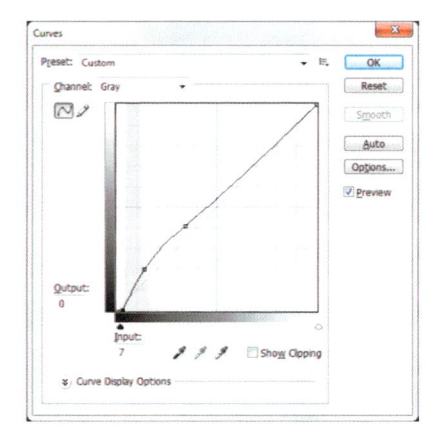

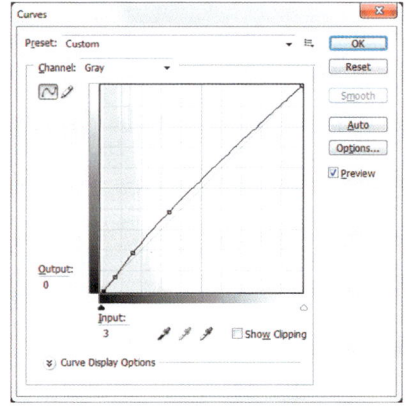

Each image needs to be stretched carefully. On the left is the first curve applied to the OIII image. The black point was set to the very left edge of the histogram, the middle anchor point was based on the nebulosity, and the top anchor is there to hold down the top part of the curve. This was the first of several curves applied to the OIII image. Similar curves were applied to the SII image.

On the right however, is the first and only stretch needed for the H-alpha image before combining color channels. This was a contrast curve, with the middle two anchor points set based on the background and dimmest nebulosity.

After stretching, the three images were combined into one color image by creating a new RGB mode image of the same dimensions, then pasting each image into the appropriate color channel. In this case we're using the HST palette, so SII was pasted into the red channel, H-alpha to green, and OIII to blue.

The gradient in the red (SII) channel is plainly apparent, and it's also clear that we'll need to do some additional stretching and color balancing.

The gradient can be removed by either using a Photoshop plug-in or switching to another program like PixInsight to perform the task. The image is also cropped appropriately to leave off the non-overlapping area from calibration. The result after gradient removal and cropping is shown at left.

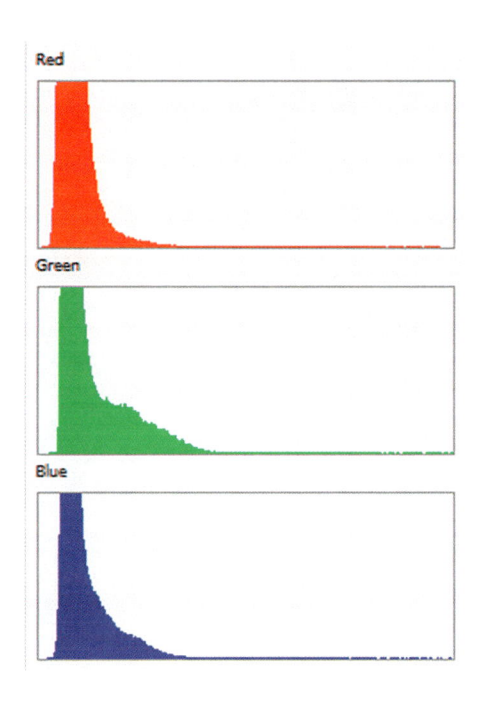

A look at the channel histograms shows why the nebula appears blue-green. While the background is balanced (the main "hump"), there is simply more nebulosity (the "toe") brightness in the H-alpha and OIII channels.

In a few steps, we'll use the Selective Color tool to emphasize the color differences, but we need better balance between the channels to give us something to work with. It's clear that the red and blue channels need to be pushed a little harder.

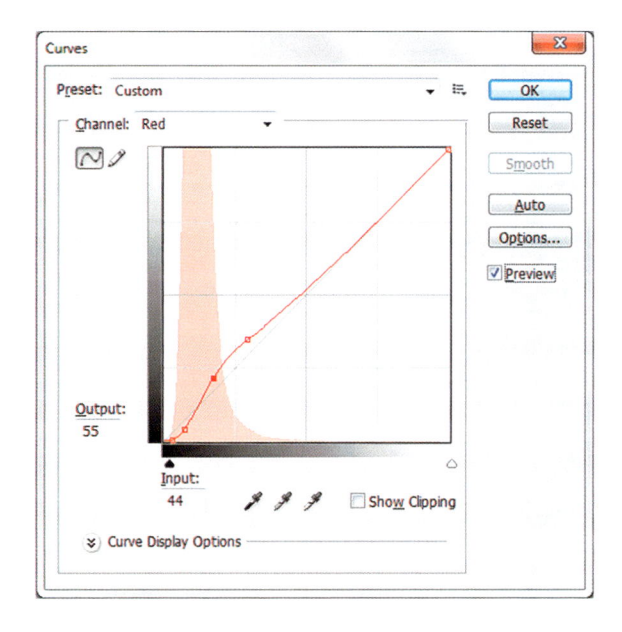

The biggest change is required in the red channel. We need to bring it up substantially in the nebula without creating a reddish background. This is done with the curve at left.

Brightening the red channel in the nebulosity will cause the areas where H-alpha and SII combine to shift from green toward yellow, and areas dominated more by SII (only on the outer fringes of the object) to show as orange or red.

The green and blue channels are also indistinct, so a similar curve is applied to the blue channel to separate the fairly uniform teal into a gradient from blue to teal to green (curve not shown).

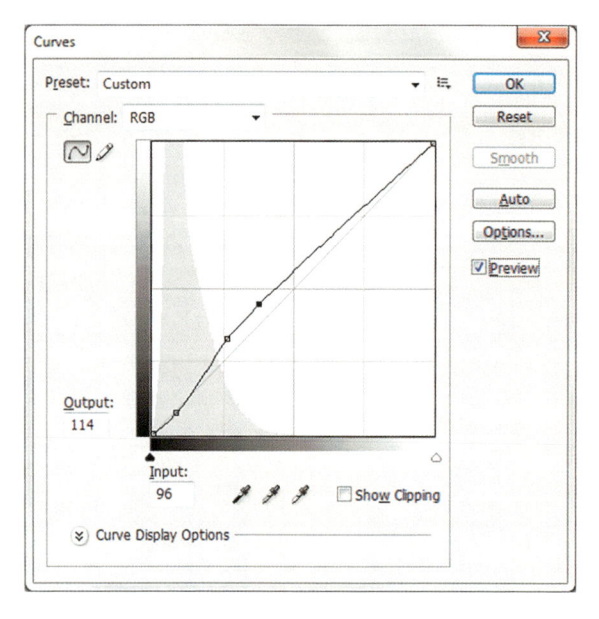

Now that the channels are more balanced across the nebula, we can apply another stretch to all channels at once. There is no point in expanding the dynamic range of the background, so the first anchor creates a subtle S-shape to compress the background a little. That anchor is placed so that the curve crosses the reference line right at the level of the dimmest nebulosity, since we want to brighten everything above those levels. The top two anchors determine how much brighter the nebula becomes and control the brightness of stars, respectively.

The result after these curves reveals a differentiated view of the emission lines in the nebula. It is both brighter and more colorful, but we'll now start working with luminance and color separately.

Processing Images

To isolate the luminosity, duplicate the image, flatten the layers, change to Lab mode, and select the Luminosity channel. Copy this image and paste it as a new layer in the original image, with the layer blending mode set to Luminosity. (The Lab mode image is a temporary step, so it can be deleted.)

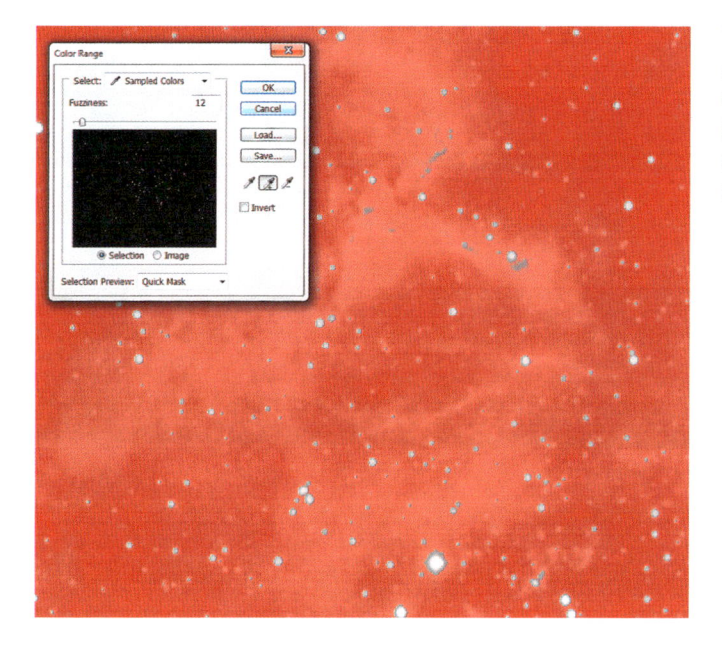

With the new luminosity layer selected, we will again select the stars using the same method we used for the OIII channel alone, including expanding and feathering the selection. While the dim stars are harder to target, this will at least select the stars brighter than the nebulosity.

Once the stars are selected, we need to put them in their own layer using the New Layer via Cut menu option. This leaves us with two layers of luminosity data, one with the brighter stars only, the other with the background, nebulosity, and dimmer stars—with holes where the brighter stars were. (Both layers are set to Luminosity layer blending mode.)

Processing the stars layer is simple: apply the Minimize filter with a radius of one pixel. The other layer is more complex. We want to sharpen this layer to emphasize the overall structure of the nebula, but also the smaller details. The requires us to copy the layer to apply these adjustments separately, so we now have a total of three luminosity layers. To one of them, the Unsharp Mask filter is applied with a radius of 50 pixels to emphasize the gross structure of the nebula. To the other, which is placed on top, the Unsharp Mask filter is also applied, but at a radius of about 2.5 pixels to emphasize the tiny crack-like features. The smaller radius setting creates too much contrast in the stars that remain in the layer, so a hide-all layer mask is applied, allowing us to selectively reveal only these thin structures with the paintbrush. The combination of these three luminosity layers is shown above.

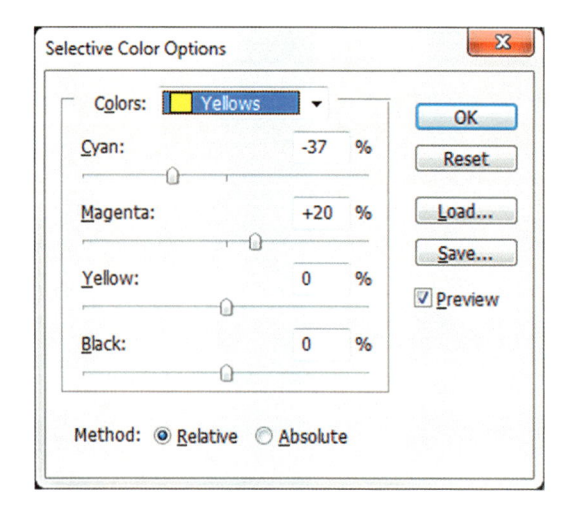

Now that the luminance is sharp, it's time for the color adjustments via the Selective Color tool. To bring the yellows toward orange, about 40% of the cyan is removed, and magenta is boosted by about 20%. (All values were determined by adjusting the image to taste; they are not empirically derived, and they are different for each image.) To bring the greens toward yellow, about 80% of the cyans and 60% of the magenta is removed from the greens. Finally, to bring the cyans toward blue, about 80% of the yellow is removed from them.

Processing Images

Our image now features sharp details and high contrast, both in the luminosity and between colors. We're almost done, but a look at the deep red SII areas around the outskirts of the nebula reveal substantial noise. Even within the brighter areas, there is color noise and some magenta haloing around stars.

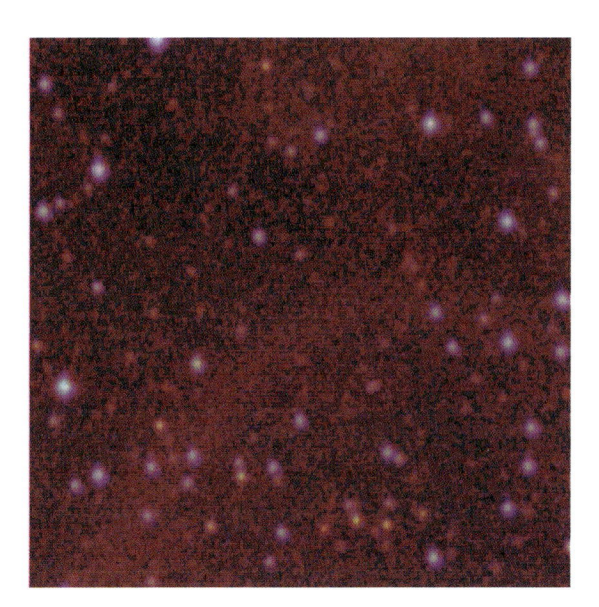

Because we have separated the luminance and color information into separate layers, we can apply noise reduction appropriately to each. The base layers where we have our color information can withstand aggressive noise reduction, especially via the Color Noise slider. Normally, applying high levels of noise reduction creates a overly smoothed or blurry look, and that's exactly what it does to this layer, but since the luminance is controlled by the layers above it, smoothing the colors reduces the appearance of color noise without affecting the overall sharpness of the image. It also eliminates magenta halos around the brighter stars.

Reducing noise in the luminance data is a bit more difficult, as it results in apparent blurring at even moderate strengths. We want to avoid this in the main nebulosity, however the background and dim nebulosity levels (especially the SII clouds) could use some serious smoothing. We can confine the effect to the dimmest regions by creating a new layer, applying a strong noise reduction, then adding a layer mask to block the brighter levels. (Yes, there are now four luminosity layers!)

The mask can be created manually via the paintbrush, or the Threshold tool (Image > Adjustments > Threshold) will help automate the task a little. The net result (before and after) is shown above on a detail from the lower-left of the image.

After a final round of very subtle curves adjustments, the final image is shown below.

A Exercise Answers

1.1 If a sensor's dark signal has a 7° C doubling temperature, and the ambient temperature is 20° C, by what percentage is the dark signal reduced when the sensor is cooled to -20° C? By what percentage is the dark noise reduced?

This 40° C reduction in temperature is 40/7=5.71 times the doubling temperature, thus the dark signal is reduced to $0.5^{5.7}$=0.019 of its original value, or a **98.1%** reduction.

The dark noise is the square root of the dark signal, so the reduction in noise is *smaller* than the reduction in signal. In this case, the square root of 0.019 is 0.138, thus the reduction in noise is about **86.2%**.

Consider an example exposure where the dark signal at 20° C is 1,000 electrons. In this case the noise contributed by this dark signal is the square root of 1,000=31.62 electrons. At -20° C, the dark signal is now only 19 electrons, so the noise is now the square root of 19=4.36 electrons. The reduction from 31.62 to 4.36 is an 86% reduction.

1.2 The KAF-8300 sensor specifications state that it has a thermal signal of no more than 200 electrons per second at 60° C, a doubling temperature of 5.8° C, and a full well capacity of 25,500 electrons. What is the thermal signal at 20° C? At that temperature how long would it take for the well to fill with thermal signal? (Assume a bias signal of zero.)

Again, we have a 40° C drop in temperature. Here, this means that the dark signal is reduced to $0.5^{20/5.8}$=0.0916 of the original level, or about 18 electrons per second. 25,500/18.3=1392, so the well would be full of dark signal electrons in 1,392 seconds, or about 23 minutes, at 20° C. Obviously, this would make long-exposure imaging difficult.

1.3 Assuming this same sensor has a read noise of 10 electrons, what is the available dynamic range for a 20-minute exposure at 20° C? And at -15° C?

185

An exposure of 20 minutes would produce 20×60×18.3=21,988 electrons. This leaves only about 3,500 electrons of capacity in the well. That's only about 14% of the well capacity available for electrons generated by light from the sky! This means the possible dynamic range—the ratio of the brightest capturable level from the sky to the dimmest—is a ratio of 3,500/148=23.6, or about 27.5 dB. This is nearly useless for imaging, as we haven't even accounted for shot noise from the target object.

At -15° C, the dark signal is $0.5^{75/5.8} \times 200 = 0.0256$ electrons per second. For a 20-minute exposure, this would generate only 20×60×0.0256=31.7 electrons on average. Here, the read noise is a meaningful addition of 10 more electrons for a total of about 42 per exposure on average. 99.9% of the well capacity is available for light from the sky now. The dynamic range would be about 25,468/42 for a ratio of about 606, or 55.7dB.

The calculations of decibels and dynamc range are not as important as highlighting that dark signal is not only a source of noise, but it also reduces the available dynamic range. Both effects should make clear the importance of cooling.

1.4 For the example light frame shown, what is the overall SNR?

There were 586 total electrons in the well, and we calculated the total noise as 23.2, so the SNR is simply 586/23.2=25.3.

1.5 Considering only the *target* signal, what is the SNR for this exposure?

Only 200 of the 586 total electrons were from the target signal, thus the SNR relative to the signal we care about is only 200/23.2=8.6.

1.6 What would the SNR be after averaging 10 exposures? How does that compare to the SNR of a single exposure that is 10 times as long (assume the three signals scale linearly)?

From 10 exposures, we'd expect to count about 586×10=5,860 total electrons. Within this, we'd expect about 522×10=5,220 to be from the signals and 64×10=640 to be from the bias current. The bias adds 15 electrons per exposure of noise, for a total of 150. The shot noise from the signals is the square root of 5,220=72.2. Summing in quadrature yields the square root of (5,220+150)=73.3 for the total noise. Thus the SNR is 5,860/73.3=80.

A single exposure 10 times as long would still be expected to collect around 5,220 electrons from the signals, but only 64 from the bias current, for a total of 5,284. The total noise would be the square root of (5,220+15)=72.4, for an SNR of 73.

Wait... why did the SNR go *down* when we took fewer shots? It's because the electron count was inflated with useless bias current when we took many exposures, making the "signal" numerator larger. This is a great demonstration that the gross SNR number is not useful. What matters is the ratio of the *signal you're interested in*, the target signal, to the noise. In that case, the 10-exposure SNR is more like 2,000/73.3=27.3, and the single long exposure is a small improvement at 2,000/72.4=27.6.

Exercise Answers

2.1 Using a refractor with a 100 mm diameter objective under skies with two-arc second seeing, what is the smallest approximate photosite size if the refractor is f/5 and you want to sample no more than twice the seeing? What if it is f/10?

The f/5 refractor would have a focal length of 500 mm. We want an image scale of 1 arc second per pixel. Plugging into the given equation,

$$Image\ scale \cong \frac{206.265 \times pixel\ size}{f}$$

and rearranging to solve for pixel size, we get 2.4 µm. This is smaller than almost any commonly used sensor's photosites. At 500 mm, it's hard to oversample unless seeing is exceptional. This is why DSLRs, with their small photosites, are great matches for widefield imaging with small refractors.

An f/10 refractor would have a focal length of 1000 mm, thus the smallest pixels to avoid oversampling at the Nyquist ratio would be 4.8 µm.

2.2 Assuming that we are imaging in visible light from 400–700 nm in wavelength, what is the smallest diameter aperture that will have better than one arc second resolution across the visible spectrum?

Since resolution improves at shorter wavelengths, we use 700 nm for the worst-case resolution. Rayleigh's Criterion (in radians) is

$$Angular\ resolution \cong \sin^{-1}(1.22\frac{\lambda}{a})$$

Setting the angular resolution to 1 arc second (=1/206265 radians) and the wavelength to 700 nm, then rearranging to solve for the aperture yields about 173 mm. (Using Dawes' Limit yields about 116 mm.)

This tells us that with average seeing or better and optics smaller than about 170 mm, aperture becomes the resolution-limiting factor before the atmosphere does.

2.3 The primary reflection halo around Alnitak in Figure 70 is 110 pixels across, and the imaging camera had pixels that were 5.4 µm square. The telescope was f/5.3. How far from the sensor was the offending reflective surface?

The light traveled an additional 110 × 5.4 µm × 5.3 = 3.15 mm, thus the reflective surfaces were about 1.6 mm apart. We can't actually say where the reflective surface is relative to the sensor, only the spacing that resulted in the halo. But assuming the system only have one additional reflective surface above the sensor, this would be the distance to that surface.

3.1 What is the color of an image after a curve is applied that consists of a straight horizontal line at the 8-bit value of 200?

Gray, at a brightness of 200. All color would be removed if this were applied to all channels in a color image.

3.2 Does using the Levels tool instead of the Curves tool to stretch an image prevent the loss of color fidelity and saturation?

No. Any reallocation of dynamic range, which is the reason for stretching, requires compressing one range in order to expand another. As we've seen, even linear adjustments using only the black and white point sliders will alter colors.

B Moonless hours

For deep-sky photography, the moon can be the ultimate light pollution. For most objects, unless you are using narrowband filters, it is best to shoot when the moon is absent from the sky. This appendix contains tables that show the number of moonless hours between astronomical twilight in a given calendar day at 40, 20, and 0 degrees north and south latitude for 2013 and 2014. These are to provide a rough guide to when the sky will be dark (and maybe when you plan your vacation to a dark sky site).

The tables do not take into account the phase of the moon. While a crescent moon may be tolerable in many situations, given that astronomical twilight occurs when the sun is more that 12 degrees below the horizon, and a quarter or smaller moon is within about 22.5 degrees of the sun, these tables should still provide a reasonable guide for most purposes.

Note that a waning moon rises after sunset, so the best imaging time is earlier for a waning moon. A waxing moon sets before sunrise, making the best imaging hours late.

2013 Tables

40 degrees N latitude   2013   Moonless hours between astronomical twilights

	1	2	3	4	5	6	7	8	9	10	11	12	13	14	15	16	17	18	19	20	21	22	23	24	25	26	27	28	29	30	31
Jan	2.6	3.7	4.7	5.6	5.8	6.9	8.0	9.1	10.3	11.3	11.2	11.2	10.4	9.3	8.2	7.1	6.1	5.7	5.1	4.1	3.2	2.3	1.4	0.7	0.0	0.0	0.0	0.1	1.1	2.2	3.2
Feb	4.3	5.1	5.4	6.5	7.6	8.6	9.5	10.3	10.5	10.4	10.2	9.1	8.2	7.0	6.0	5.4	5.0	4.1	3.2	2.4	1.6	0.9	0.3	0.0	0.0	0.0	0.6	1.7			
Mar	2.8	3.9	4.6	5.0	6.0	6.9	7.7	8.5	9.1	9.3	9.3	8.6	7.6	6.6	5.6	4.7	4.6	3.8	3.0	2.3	1.6	1.0	0.5	0.0	0.0	0.0	0.0	1.2	2.3	3.4
Apr	4.1	4.3	5.2	5.9	6.6	7.1	7.7	7.9	7.8	7.8	7.7	7.0	6.0	5.1	4.3	3.7	3.6	2.9	2.3	1.7	1.1	0.6	0.0	0.0	0.0	0.0	0.5	1.6	2.5	3.3	
May	3.4	3.9	4.5	5.1	5.6	6.0	6.3	6.3	6.2	6.2	6.1	5.6	4.9	4.2	3.5	3.0	2.9	2.4	1.9	1.3	0.8	0.2	0.0	0.0	0.0	0.5	1.3	1.9	2.5	2.7	
Jun	3.0	3.5	4.0	4.5	5.0	5.1	5.0	5.0	5.0	5.0	4.5	4.0	3.5	3.0	2.5	2.0	1.4	0.7	0.0	0.0	0.0	0.0	0.0	0.0	0.2	0.8	1.3	1.9	2.4		
Jul	2.4	2.9	3.5	4.2	4.9	5.1	5.1	5.2	5.2	5.2	5.2	4.7	4.2	3.7	3.1	2.8	2.4	1.6	0.7	0.0	0.0	0.0	0.0	0.0	0.3	0.8	1.4	2.0	2.6	3.0	
Aug	3.4	4.2	5.0	5.9	6.3	6.4	6.4	6.5	6.5	6.6	6.3	5.7	5.1	4.3	3.5	3.5	2.5	1.5	0.3	0.0	0.0	0.0	0.1	0.7	1.4	2.1	2.9	3.7	3.8	4.6	
Sep	5.5	6.5	7.5	7.9	8.0	8.2	8.1	7.6	6.9	6.1	5.2	4.2	4.1	3.1	2.0	0.8	0.0	0.0	0.0	0.0	0.2	0.9	1.6	2.4	3.2	4.2	4.7	5.1	6.1		
Oct	7.1	8.1	9.1	9.4	9.5	9.4	9.4	8.7	7.8	6.8	5.7	4.6	4.6	3.5	2.4	1.3	0.2	0.0	0.0	0.0	0.2	1.0	1.8	2.7	3.6	4.5	5.4	5.5	6.5	7.5	
Nov	9.6	10.5	10.5	10.6	10.4	9.4	8.3	7.2	6.1	5.1	5.0	3.9	2.8	1.8	0.7	0.0	0.0	0.0	0.1	0.9	1.8	2.7	3.7	4.6	5.6	5.8	6.6	7.6	8.6	9.7	
Dec	10.9	11.3	11.3	11.0	9.9	8.7	7.5	6.4	5.6	5.4	4.3	3.3	2.3	1.3	0.3	0.0	0.0	0.0	0.6	1.5	2.4	3.4	4.3	5.3	5.7	6.3	7.3	8.4	9.5	10.6	11.4

20 degrees N latitude   2013   Moonless hours between astronomical twilights

	1	2	3	4	5	6	7	8	9	10	11	12	13	14	15	16	17	18	19	20	21	22	23	24	25	26	27	28	29	30	31
Jan	2.4	3.3	4.1	5.1	5.1	6.0	7.0	8.0	9.1	10.1	10.4	10.4	9.7	8.7	7.8	6.9	6.0	5.4	5.2	4.3	3.5	2.7	1.8	1.0	0.3	0.0	0.0	0.1	0.9	1.8	2.7
Feb	3.7	4.6	4.9	5.6	6.6	7.6	8.6	9.5	10.1	10.1	10.0	9.1	8.2	7.3	6.4	5.5	5.2	4.7	3.8	3.0	2.2	1.5	0.7	0.0	0.0	0.0	0.5	1.4			
Mar	2.4	3.4	4.4	4.7	5.4	6.3	7.2	8.1	8.9	9.6	9.6	9.6	9.2	8.3	7.4	6.5	5.7	4.8	4.8	4.0	3.2	2.5	1.8	1.1	0.4	0.0	0.0	0.1	1.1	2.1	3.1
Apr	4.1	4.5	5.0	5.9	6.7	7.4	8.1	8.8	9.0	9.0	9.0	8.2	7.4	6.5	5.7	4.9	4.4	4.2	3.5	2.8	2.1	1.4	0.7	0.0	0.0	0.0	0.7	1.7	2.7	3.6	
May	4.3	4.4	5.2	5.9	6.6	7.2	7.8	8.4	8.4	8.3	8.3	7.4	6.7	5.9	5.2	4.5	4.1	3.9	3.2	2.6	1.9	1.1	0.3	0.0	0.0	0.2	1.2	2.1	2.9	3.6	4.0
Jun	4.3	5.0	5.6	6.3	6.9	7.6	7.9	7.9	7.9	7.8	7.1	6.4	5.8	5.1	4.5	4.0	3.9	3.2	2.5	1.7	0.8	0.0	0.0	0.0	0.6	1.4	2.1	2.8	3.5	3.9	
Jul	4.2	4.8	5.5	6.2	7.0	7.8	7.9	7.9	8.0	7.9	7.3	6.6	6.0	5.4	4.7	4.1	4.0	3.1	2.2	1.2	0.2	0.0	0.0	0.1	0.8	1.5	2.2	2.9	3.6	4.0	4.4
Aug	5.1	5.9	6.7	7.6	8.4	8.4	8.5	8.5	8.2	7.6	6.9	6.2	5.4	4.6	4.4	3.7	2.7	1.7	0.6	0.0	0.0	0.0	0.5	1.2	1.9	2.7	3.4	4.2	4.5	5.0	5.9
Sep	6.7	7.6	8.3	9.1	9.1	9.1	9.1	8.4	7.6	6.8	5.9	5.0	4.6	4.0	3.0	1.9	0.9	0.0	0.0	0.0	0.2	1.0	1.8	2.5	3.3	4.2	4.9	5.0	5.9	6.7	
Oct	7.6	8.5	9.3	9.7	9.7	9.7	9.0	8.1	7.2	6.2	5.2	4.7	4.2	3.2	2.2	1.3	0.3	0.0	0.0	0.0	0.8	1.6	2.4	3.2	4.0	4.9	5.3	5.7	6.6	7.4	8.3
Nov	9.2	10.1	10.1	10.2	9.5	8.5	7.4	6.4	5.4	4.8	4.5	3.5	2.6	1.7	0.8	0.0	0.0	0.0	0.4	1.2	2.1	2.9	3.7	4.5	5.4	5.4	6.2	7.1	8.0	8.9	
Dec	9.9	10.4	10.4	9.9	8.8	7.8	6.8	5.8	5.1	4.9	4.0	3.1	2.2	1.4	0.5	0.0	0.0	0.0	0.7	1.5	2.3	3.1	3.9	4.7	5.2	5.6	6.5	7.4	8.4	9.4	10.4

0 degrees latitude   2013   Moonless hours between astronomical twilights

	1	2	3	4	5	6	7	8	9	10	11	12	13	14	15	16	17	18	19	20	21	22	23	24	25	26	27	28	29	30	31
Jan	2.0	2.8	3.5	4.3	4.6	5.2	6.1	7.0	8.0	9.1	9.4	9.4	9.0	8.1	7.3	6.5	5.8	5.0	4.9	4.3	3.5	2.8	2.0	1.2	0.4	0.0	0.0	0.0	0.6	1.4	2.2
Feb	3.0	3.9	4.5	4.8	5.8	6.8	7.8	8.8	9.5	9.5	9.5	8.9	8.1	7.3	6.5	5.7	5.0	4.4	4.1	3.3	2.5	1.7	1.0	0.2	0.0	0.0	0.2	1.0			
Mar	1.9	2.8	3.8	4.6	4.7	5.7	6.7	7.6	8.5	9.4	9.6	9.6	9.4	8.6	7.8	7.0	6.2	5.4	4.9	4.6	3.8	3.0	2.2	1.4	0.7	0.0	0.0	0.0	0.8	1.8	2.8
Apr	3.8	4.7	4.8	5.7	6.6	7.4	8.3	9.1	9.6	9.6	9.6	9.0	8.2	7.4	6.6	5.8	5.0	4.8	4.2	3.4	2.7	1.9	1.1	0.2	0.0	0.0	0.7	1.7	2.7	3.7	
May	4.6	4.8	5.5	6.3	7.1	7.8	8.6	9.4	9.5	9.5	9.4	8.5	7.8	7.0	6.2	5.4	4.7	4.7	4.0	3.2	2.4	1.5	0.6	0.0	0.0	0.4	1.4	2.4	3.3	4.2	4.7
Jun	5.0	5.7	6.5	7.3	8.0	8.8	9.4	9.4	9.4	9.0	8.3	7.5	6.8	6.1	5.3	4.7	4.5	3.8	2.9	2.0	1.0	0.0	0.0	0.0	1.0	1.9	2.8	3.6	4.4	4.6	
Jul	5.1	5.9	6.7	7.5	8.3	9.1	9.4	9.4	9.4	8.9	8.2	7.4	6.7	5.9	5.1	4.8	4.3	3.4	2.4	1.4	0.3	0.0	0.0	0.5	1.4	2.2	3.0	3.8	4.6	4.6	5.4
Aug	6.2	7.0	7.8	8.6	9.4	9.5	9.5	9.5	8.8	7.2	6.4	5.5	4.8	4.6	3.6	2.6	1.6	0.6	0.0	0.0	0.2	1.0	1.8	2.6	3.4	4.3	4.8	5.1	5.9	6.7	
Sep	7.5	8.2	9.0	9.6	9.6	9.6	9.2	8.3	7.5	6.6	5.6	4.7	4.6	3.6	2.7	1.7	0.8	0.0	0.0	0.0	0.6	1.4	2.3	3.1	3.9	4.7	5.0	5.5	6.3	7.1	
Oct	7.8	8.6	9.3	9.6	9.6	9.4	8.5	7.5	6.6	5.6	4.6	4.6	3.7	2.8	1.9	1.1	0.3	0.0	0.0	0.2	1.0	1.9	2.7	3.5	4.2	5.0	5.4	5.8	6.5	7.2	8.0
Nov	8.8	9.5	9.5	9.5	8.6	7.6	6.6	5.7	4.8	4.5	3.9	3.1	2.3	1.5	0.7	0.0	0.0	0.0	0.5	1.3	2.1	2.8	3.6	4.3	4.9	5.0	5.7	6.5	7.3	8.1	
Dec	9.0	9.4	9.4	8.9	7.9	7.0	6.0	5.2	4.6	4.4	3.6	2.8	2.0	1.2	0.4	0.0	0.0	0.0	0.5	1.3	2.0	2.7	3.4	4.2	4.7	4.9	5.7	6.5	7.4	8.4	9.4

Moonlight Calendars

20 degrees S latitude   **2013**   Moonless hours between astronomical twilights

	1	2	3	4	5	6	7	8	9	10	11	12	13	14	15	16	17	18	19	20	21	22	23	24	25	26	27	28	29	30	31
Jan	1.4	2.1	2.7	3.4	3.9	4.1	4.9	5.8	6.8	7.8	8.0	8.0	8.0	7.3	6.6	6.0	5.4	4.7	4.3	4.1	3.4	2.7	1.9	1.1	0.3	0.0	0.0	0.0	0.1	0.8	1.5
Feb	2.2	3.0	3.8	4.0	4.7	5.7	6.8	7.8	8.6	8.6	8.6	8.5	7.8	7.2	6.5	5.8	5.1	4.6	4.3	3.5	2.7	1.9	1.0	0.1	0.0	0.0	0.0	0.5			
Mar	1.3	2.1	3.0	4.0	4.4	5.0	6.0	7.0	8.1	9.1	9.3	9.3	9.3	8.8	8.1	7.4	6.6	5.8	5.0	4.9	4.2	3.4	2.5	1.6	0.7	0.0	0.0	0.0	0.4	1.3	2.3
Apr	3.3	4.3	4.8	5.3	6.4	7.3	8.3	9.2	9.8	9.8	9.9	9.6	8.9	8.1	7.3	6.4	5.6	5.0	4.8	3.9	3.1	2.2	1.3	0.3	0.0	0.0	0.4	1.4	2.5	3.5	
May	4.5	5.1	5.5	6.5	7.4	8.3	9.2	10.0	10.2	10.2	10.3	9.5	8.7	7.8	7.0	6.2	5.3	5.1	4.5	3.6	2.7	1.7	0.7	0.0	0.0	0.3	1.4	2.5	3.5	4.5	5.2
Jun	5.4	6.3	7.2	8.1	8.9	9.8	10.4	10.4	10.4	10.1	9.2	8.4	7.6	6.7	5.9	5.2	5.0	4.1	3.1	2.1	1.1	0.0	0.0	0.1	1.2	2.2	3.2	4.1	5.0	5.2	
Jul	5.9	6.8	7.6	8.5	9.3	10.1	10.4	10.4	10.4	9.7	8.9	8.0	7.2	6.3	5.3	5.3	4.4	3.4	2.4	1.3	0.3	0.0	0.0	0.8	1.8	2.7	3.6	4.5	5.0	5.4	6.3
Aug	7.1	7.9	8.7	9.5	10.2	10.2	10.2	9.9	9.1	8.2	7.3	6.4	5.4	5.1	4.4	3.4	2.4	1.4	0.5	0.0	0.0	0.4	1.3	2.2	3.2	4.0	4.9	4.9	5.7	6.5	7.3
Sep	8.0	8.7	9.4	9.8	9.9	9.8	9.0	8.1	7.1	6.1	5.1	4.8	4.1	3.2	2.3	1.4	0.6	0.0	0.0	0.0	0.9	1.8	2.6	3.5	4.3	4.8	5.1	5.8	6.6	7.2	
Oct	7.9	8.5	9.2	9.2	9.2	8.8	7.8	6.8	5.8	4.8	4.3	3.9	3.0	2.2	1.5	0.7	0.0	0.0	0.0	0.3	1.2	2.0	2.8	3.6	4.3	4.6	4.9	5.6	6.2	6.9	7.5
Nov	8.2	8.6	8.5	8.5	7.6	6.6	5.6	4.7	3.9	3.9	3.2	2.4	1.7	1.0	0.3	0.0	0.0	0.0	0.4	1.1	1.9	2.6	3.2	3.8	4.3	4.4	5.1	5.7	6.4	7.1	
Dec	7.9	8.0	8.0	7.7	6.8	5.9	5.1	4.3	3.8	3.7	3.0	2.3	1.6	0.9	0.1	0.0	0.0	0.0	0.2	0.9	1.5	2.1	2.7	3.4	3.9	4.0	4.7	5.4	6.2	7.2	7.9

40 degrees S latitude   **2013**   Moonless hours between astronomical twilights

	1	2	3	4	5	6	7	8	9	10	11	12	13	14	15	16	17	18	19	20	21	22	23	24	25	26	27	28	29	30	31
Jan	0.2	0.7	1.2	1.7	2.3	2.5	3.0	3.8	4.7	5.3	5.3	5.3	5.4	5.4	5.4	4.9	4.4	3.9	3.4	3.0	2.8	2.2	1.4	0.6	0.0	0.0	0.0	0.0	0.0	0.0	0.2
Feb	0.8	1.5	2.2	3.0	3.1	4.0	5.1	6.3	6.8	6.8	6.9	6.9	7.0	6.7	6.2	5.6	4.9	4.3	3.9	3.5	2.7	1.8	0.8	0.0	0.0	0.0	0.0	0.0			
Mar	0.3	1.0	1.8	2.7	3.7	3.9	4.9	6.0	7.2	8.0	8.4	8.4	8.5	8.1	7.6	7.0	6.2	5.4	4.5	4.5	3.6	2.7	1.7	0.7	0.0	0.0	0.0	0.0	0.0	0.5	1.4
Apr	2.4	3.5	4.7	4.7	5.8	7.0	8.1	9.2	9.7	9.8	9.8	9.9	9.5	8.8	8.0	7.2	6.3	5.3	5.0	4.4	3.4	2.3	1.3	0.2	0.0	0.0	0.0	0.9	2.0	3.1	
May	4.3	5.4	5.4	6.5	7.6	8.7	9.7	10.7	10.8	10.8	10.8	10.5	9.7	8.8	7.9	6.9	6.0	5.5	5.0	4.0	2.9	1.8	0.6	0.0	0.0	0.0	1.2	2.4	3.6	4.7	5.7
Jun	5.8	6.8	7.9	8.9	9.9	10.8	11.4	11.4	11.4	11.2	10.2	9.3	8.4	7.4	6.4	5.7	5.4	4.3	3.2	2.1	0.9	0.0	0.0	0.0	1.2	2.4	3.5	4.6	5.6	5.7	
Jul	6.7	7.7	8.7	9.6	10.5	11.3	11.3	11.3	11.3	10.5	9.5	8.5	7.5	6.5	5.7	5.4	4.3	3.2	2.1	1.0	0.0	0.0	0.0	0.9	2.1	3.2	4.2	5.3	5.4	6.2	7.2
Aug	8.1	8.9	9.7	10.4	10.7	10.7	10.7	10.3	9.3	8.3	7.2	6.1	5.3	5.0	3.9	2.8	1.8	0.9	0.1	0.0	0.0	0.5	1.6	2.6	3.7	4.6	5.0	5.5	6.4	7.2	7.9
Sep	8.6	9.2	9.7	9.7	9.6	9.6	8.7	7.6	6.5	5.4	4.6	4.3	3.3	2.3	1.5	0.8	0.1	0.0	0.0	0.0	1.0	2.0	2.9	3.8	4.5	4.6	5.4	6.1	6.7	7.2	
Oct	7.8	8.3	8.3	8.3	7.9	6.7	5.6	4.6	3.8	3.6	2.7	2.0	1.3	0.7	0.1	0.0	0.0	0.0	0.1	1.0	1.9	2.7	3.4	3.8	4.0	4.6	5.1	5.6	6.1	6.6	
Nov	6.8	6.7	6.6	6.6	6.0	4.9	4.0	3.2	2.9	2.5	1.9	1.3	0.7	0.1	0.0	0.0	0.0	0.0	0.0	0.6	1.2	1.8	2.4	2.9	3.0	3.3	3.8	4.3	4.8	5.3	
Dec	5.3	5.2	5.2	5.2	4.8	4.0	3.4	2.8	2.4	2.2	1.7	1.1	0.6	0.0	0.0	0.0	0.0	0.0	0.0	0.4	0.9	1.3	1.8	2.3	2.4	2.9	3.5	4.2	5.0	5.0	

2014 Tables

40 degrees N latitude   **2014**   Moonless hours between astronomical twilights

	1	2	3	4	5	6	7	8	9	10	11	12	13	14	15	16	17	18	19	20	21	22	23	24	25	26	27	28	29	30	31
Jan	11.4	11.3	10.1	8.9	7.8	6.6	5.8	5.6	4.5	3.5	2.6	1.7	0.8	0.0	0.0	0.0	0.0	0.9	1.8	2.8	3.7	4.7	5.3	5.7	6.8	7.8	8.9	9.8	10.8	10.8	10.8
Feb	10.0	8.8	7.6	6.5	5.6	5.5	4.5	3.5	2.6	1.8	1.0	0.4	0.0	0.0	0.0	0.2	1.2	2.1	3.1	4.1	4.8	5.2	6.2	7.1	8.1	8.9	9.7	9.7			
Mar	9.7	9.5	8.3	7.2	6.1	5.1	4.9	4.2	3.3	2.5	1.8	1.1	0.5	0.0	0.0	0.0	0.0	0.5	1.5	2.6	3.6	4.3	4.5	5.5	6.3	7.0	7.7	8.3	8.4	8.3	8.1
Apr	7.7	6.6	5.6	4.7	4.1	3.8	3.1	2.4	1.8	1.2	0.6	0.1	0.0	0.0	0.0	0.0	0.0	0.9	1.9	2.8	3.6	3.7	4.4	5.1	5.7	6.3	6.9	6.9	6.8	6.8	
May	6.1	5.2	4.4	3.7	3.2	3.0	2.4	1.8	1.3	0.8	0.3	0.0	0.0	0.0	0.0	0.0	0.9	1.8	2.5	2.9	3.1	3.7	4.2	4.8	5.3	5.5	5.4	5.4	5.3	5.3	5.1
Jun	4.4	3.8	3.3	2.8	2.5	2.3	1.8	1.2	0.7	0.1	0.0	0.0	0.0	0.0	0.0	0.7	1.3	1.9	2.4	2.5	3.0	3.6	4.2	4.9	4.9	4.9	4.9	4.9	4.9	5.0	
Jul	4.7	4.3	3.8	3.3	2.8	2.6	2.3	1.6	0.9	0.1	0.0	0.0	0.0	0.0	0.0	0.7	1.3	1.9	2.5	2.7	3.2	3.9	4.7	5.6	5.8	5.8	5.9	5.9	6.0	6.1	5.8
Aug	5.3	4.8	4.2	3.6	3.3	2.9	2.0	1.0	0.0	0.0	0.0	0.0	0.0	0.6	1.2	1.9	2.6	3.4	3.5	4.3	5.1	6.0	7.0	7.3	7.4	7.4	7.5	7.6	7.4	6.8	6.2
Sep	5.5	4.7	3.9	3.8	2.8	1.7	0.5	0.0	0.0	0.0	0.0	0.7	1.4	2.2	3.0	3.9	4.4	4.8	5.7	6.7	7.6	8.6	8.9	8.9	9.0	9.0	8.7	8.1	7.3	6.4	
Oct	5.5	4.5	4.4	3.3	2.2	1.0	0.0	0.0	0.0	0.8	1.6	2.5	3.4	4.3	5.2	5.3	6.0	7.2	8.1	9.1	10.1	10.1	10.2	10.2	9.8	8.8	8.0	7.0	5.9	4.9	
Nov	4.8	3.7	2.6	1.4	0.3	0.0	0.0	0.0	0.8	1.7	2.6	3.5	4.5	5.4	5.2	6.4	7.3	8.1	9.3	10.3	11.1	11.4	11.1	10.6	9.6	8.5	7.4	6.3	5.4	5.2	
Dec	4.1	3.0	1.9	0.8	0.0	0.0	0.0	0.5	1.5	2.4	3.3	4.3	5.2	5.8	6.1	7.1	8.0	9.0	10.0	11.1	11.4	11.4	11.1	9.9	8.8	7.6	6.5	5.7	5.4	4.3	3.2

20 degrees N latitude   **2014**   Moonless hours between astronomical twilights

	1	2	3	4	5	6	7	8	9	10	11	12	13	14	15	16	17	18	19	20	21	22	23	24	25	26	27	28	29	30	31
Jan	10.4	10.3	9.3	8.2	7.3	6.3	5.4	5.3	4.5	3.6	2.7	1.9	1.1	0.3	0.0	0.0	0.0	0.8	1.6	2.4	3.2	4.1	5.0	5.0	5.9	6.9	7.9	8.9	9.9	10.2	10.2
Feb	9.6	8.6	7.6	6.6	5.7	5.3	4.8	3.9	3.1	2.3	1.5	0.8	0.1	0.0	0.0	0.2	1.0	1.8	2.7	3.6	4.5	4.7	5.5	6.5	7.4	8.4	9.2	9.8			
Mar	9.8	9.7	8.7	7.7	6.8	5.8	5.0	4.9	4.1	3.3	2.5	1.8	1.1	0.5	0.0	0.0	0.0	0.5	1.4	2.3	3.3	4.2	4.6	5.1	6.1	6.9	7.8	8.5	9.3	9.2	9.2
Apr	8.6	7.7	6.7	5.9	5.0	4.6	4.2	3.5	2.8	2.1	1.5	0.8	0.2	0.0	0.0	0.0	0.0	1.0	2.0	2.9	3.8	4.4	4.7	5.3	6.3	7.0	7.8	8.5	8.6	8.6	8.5
May	7.6	6.8	6.0	5.2	4.5	4.2	3.8	3.1	2.5	1.9	1.2	0.5	0.0	0.0	0.0	0.5	1.5	2.4	3.2	4.0	4.1	4.8	5.5	6.2	6.9	7.6	8.1	8.0	8.0	7.8	7.1
Jun	6.3	5.6	5.0	4.4	4.0	3.7	3.1	2.5	1.8	1.0	0.2	0.0	0.0	0.0	0.9	1.8	2.6	3.3	3.9	4.0	4.7	5.5	6.2	6.9	7.7	7.9	7.9	7.9	7.7	7.1	
Jul	6.4	5.8	5.2	4.6	4.0	4.0	3.3	2.5	1.7	0.8	0.0	0.0	0.0	0.4	1.2	2.0	2.7	3.5	3.9	4.2	5.0	5.8	6.6	7.4	8.2	8.2	8.2	8.3	8.1	7.5	6.8
Aug	6.2	5.6	4.9	4.3	4.1	3.3	2.3	1.3	0.3	0.0	0.0	0.1	0.9	1.7	2.5	3.2	4.0	4.3	4.9	5.7	6.5	7.4	8.2	8.8	8.8	8.9	8.9	8.4	7.8	7.1	6.4
Sep	5.6	4.7	4.5	3.8	2.8	1.7	0.7	0.0	0.0	0.0	0.7	1.6	2.4	3.2	4.0	4.8	4.9	5.8	6.6	7.4	8.3	9.1	9.5	9.5	9.5	9.2	8.5	7.7	6.9	6.0	
Oct	5.0	4.6	4.1	3.0	2.0	1.0	0.0	0.0	0.0	0.6	1.4	2.3	3.1	4.0	4.9	5.2	5.7	6.5	7.4	8.2	9.0	9.9	10.0	10.0	9.9	9.1	8.2	7.2	6.3	5.3	4.8
Nov	4.3	3.3	2.3	1.3	0.4	0.0	0.0	0.3	1.2	2.1	2.9	3.8	4.6	5.4	5.4	6.2	7.0	7.9	8.7	9.6	10.3	10.3	10.3	9.6	8.6	7.6	6.6	5.6	5.0	4.6	
Dec	3.7	2.7	1.8	0.8	0.0	0.0	0.0	0.7	1.6	2.4	3.2	4.0	4.8	5.3	5.6	6.4	7.2	8.1	9.0	10.0	10.5	10.5	10.0	9.0	8.0	7.0	6.0	5.3	5.0	4.1	3.2

0 degrees latitude   **2014**   Moonless hours between astronomical twilights

	1	2	3	4	5	6	7	8	9	10	11	12	13	14	15	16	17	18	19	20	21	22	23	24	25	26	27	28	29	30	31
Jan	9.4	9.4	8.5	7.6	6.7	5.9	5.1	4.8	4.3	3.5	2.7	1.9	1.1	0.3	0.0	0.0	0.0	0.5	1.3	2.0	2.7	3.5	4.3	4.5	5.1	6.0	7.0	8.0	9.0	9.5	9.5
Feb	9.1	8.2	7.4	6.6	5.7	5.0	4.9	4.1	3.3	2.5	1.7	0.9	0.2	0.0	0.0	0.0	0.7	1.5	2.2	3.1	3.9	4.6	4.8	5.8	6.8	7.8	8.8	9.6			
Mar	9.6	9.6	8.8	7.9	7.1	6.2	5.4	5.0	4.6	3.8	3.0	2.2	1.4	0.7	0.0	0.0	0.0	0.3	1.1	2.0	2.9	3.8	4.7	4.8	5.7	6.7	7.6	8.5	9.4	9.6	9.6
Apr	9.2	8.3	7.4	6.6	5.7	4.9	4.8	4.1	3.4	2.7	1.9	1.2	0.4	0.0	0.0	0.0	0.9	1.8	2.8	3.8	4.7	4.8	5.6	6.5	7.4	8.2	9.1	9.5	9.5	9.5	
May	8.6	7.8	6.9	6.1	5.4	4.7	4.6	3.9	3.2	2.4	1.7	0.9	0.1	0.0	0.0	0.0	1.6	2.6	3.6	4.5	4.8	5.3	6.1	7.0	7.8	8.6	9.4	9.4	9.4	9.0	8.2
Jun	7.4	6.7	6.0	5.2	4.7	4.5	3.8	3.1	2.3	1.4	0.6	0.0	0.0	0.3	1.3	2.3	3.1	4.0	4.7	4.8	5.7	6.5	7.3	8.1	9.0	9.4	9.4	9.4	8.8	8.1	
Jul	7.4	6.7	6.0	5.3	4.8	4.5	3.7	2.9	2.0	1.0	0.0	0.0	0.0	0.9	1.8	2.7	3.5	4.4	4.6	5.2	6.1	6.9	7.7	8.5	9.3	9.5	9.5	9.5	8.8	8.1	7.4
Aug	6.6	5.9	5.1	4.9	4.2	3.3	2.4	1.4	0.4	0.0	0.0	0.6	1.5	2.3	3.2	4.1	4.7	4.9	5.8	6.6	7.4	8.2	9.0	9.6	9.6	9.6	9.3	8.6	7.9	7.1	6.3
Sep	5.4	4.8	4.5	3.6	2.6	1.6	0.6	0.0	0.0	0.3	1.2	2.1	3.0	3.8	4.7	4.9	5.5	6.3	7.1	7.9	8.6	9.3	9.6	9.6	9.6	9.0	8.2	7.3	6.4	5.5	
Oct	4.6	4.6	3.6	2.7	1.7	0.8	0.0	0.0	0.0	0.9	1.8	2.6	3.5	4.3	5.0	5.1	5.9	6.6	7.4	8.1	8.8	9.5	9.6	9.6	9.2	8.4	7.4	6.5	5.6	4.7	4.5
Nov	3.7	2.9	2.0	1.1	0.2	0.0	0.0	0.4	1.3	2.2	3.0	3.8	4.5	5.0	5.2	5.9	6.7	7.4	8.1	8.9	9.4	9.4	9.4	8.6	7.7	6.7	5.8	4.9	4.5	4.1	
Dec	3.2	2.4	1.5	0.7	0.0	0.0	0.0	0.6	1.4	2.2	2.9	3.7	4.4	4.8	5.0	5.8	6.5	7.3	8.1	9.0	9.4	9.4	9.1	8.1	7.2	6.3	5.4	4.7	4.6	3.8	2.9

Moonlight Calendars

20 degrees S latitude — 2014 — Moonless hours between astronomical twilights

	1	2	3	4	5	6	7	8	9	10	11	12	13	14	15	16	17	18	19	20	21	22	23	24	25	26	27	28	29	30	31
Jan	7.9	7.9	7.5	6.7	5.9	5.2	4.5	4.1	3.9	3.2	2.4	1.7	0.9	0.1	0.0	0.0	0.0	0.1	0.7	1.3	2.0	2.6	3.3	3.9	4.1	4.9	5.9	6.9	8.0	8.3	8.4
Feb	8.4	7.7	7.0	6.3	5.6	4.8	4.5	4.1	3.3	2.5	1.7	0.9	0.1	0.0	0.0	0.0	0.2	0.9	1.6	2.3	3.1	4.0	4.2	5.0	6.0	7.0	8.1	9.0			
Mar	9.0	9.1	8.7	8.0	7.2	6.4	5.7	4.9	4.8	4.1	3.2	2.4	1.6	0.8	0.0	0.0	0.0	0.0	0.7	1.5	2.4	3.3	4.3	4.7	5.3	6.3	7.3	8.4	9.4	9.6	9.7
Apr	9.5	8.7	7.9	7.1	6.3	5.5	4.9	4.7	3.9	3.0	2.2	1.4	0.6	0.0	0.0	0.0	0.7	1.6	2.5	3.5	4.5	5.0	5.6	6.6	7.6	8.5	9.5	10.1	10.1	10.1	
May	9.4	8.6	7.7	6.9	6.1	5.3	5.1	4.5	3.6	2.8	2.0	1.1	0.2	0.0	0.6	1.6	2.6	3.6	4.6	5.2	5.6	6.6	7.6	8.5	9.4	10.3	10.4	10.4	10.0	9.2	
Jun	8.3	7.5	6.7	5.9	5.2	5.1	4.3	3.4	2.5	1.6	0.7	0.0	0.0	0.4	1.5	2.5	3.5	4.5	5.2	5.4	6.4	7.3	8.2	9.1	10.0	10.5	10.5	10.5	9.7	8.9	
Jul	8.1	7.3	6.5	5.7	5.3	4.8	3.9	3.0	2.0	1.0	0.0	0.0	0.1	1.2	2.2	3.2	4.2	5.1	5.1	6.0	6.9	7.8	8.7	9.5	10.2	10.3	10.1	9.3	8.5	7.7	
Aug	6.9	6.0	5.2	5.1	4.2	3.2	2.2	1.2	0.3	0.0	0.0	0.8	1.8	2.8	3.8	4.7	5.0	5.6	6.5	7.3	8.1	8.9	9.5	10.0	10.0	9.9	9.5	8.6	7.8	6.9	6.0
Sep	5.1	4.9	4.1	3.2	2.2	1.3	0.4	0.0	0.0	0.5	1.5	2.4	3.4	4.3	4.9	5.1	6.0	6.7	7.4	8.1	8.8	9.4	9.5	9.4	9.4	8.6	7.7	6.8	5.8	4.9	
Oct	4.5	3.9	3.0	2.2	1.3	0.5	0.0	0.0	0.0	1.0	2.0	2.8	3.7	4.5	4.7	5.2	5.9	6.6	7.2	7.8	8.4	8.8	8.8	8.4	7.5	6.5	5.6	4.7	4.0	3.8	
Nov	3.0	2.2	1.5	0.7	0.0	0.0	0.0	0.4	1.2	2.1	2.8	3.6	4.2	4.4	4.8	5.5	6.1	6.7	7.3	8.0	8.2	8.1	8.1	7.5	6.6	5.7	4.8	4.1	3.8	3.3	
Dec	2.6	1.8	1.1	0.3	0.0	0.0	0.0	0.4	1.1	1.8	2.5	3.1	3.7	4.0	4.3	4.9	5.5	6.2	7.0	7.8	7.9	7.9	7.9	7.0	6.2	5.4	4.7	4.0	4.0	3.3	2.5

40 degrees S latitude — 2014 — Moonless hours between astronomical twilights

	1	2	3	4	5	6	7	8	9	10	11	12	13	14	15	16	17	18	19	20	21	22	23	24	25	26	27	28	29	30	31
Jan	5.0	5.0	5.0	5.0	4.5	3.9	3.4	2.8	2.7	2.3	1.7	1.0	0.3	0.0	0.0	0.0	0.0	0.0	0.1	0.6	1.2	1.7	2.4	2.8	3.2	4.1	5.1	6.2	6.2	6.2	
Feb	6.3	6.4	6.1	5.6	5.0	4.4	3.7	3.6	3.1	2.3	1.5	0.6	0.0	0.0	0.0	0.0	0.0	0.5	1.1	1.9	2.7	3.6	3.6	4.7	5.8	7.1	7.8				
Mar	7.9	7.9	8.0	7.8	7.2	6.5	5.8	5.0	4.3	4.2	3.4	2.5	1.6	0.7	0.0	0.0	0.0	0.0	0.0	0.7	1.4	2.3	3.3	4.4	4.4	5.6	6.8	8.0	9.2	9.3	9.3
Apr	9.4	9.2	8.5	7.7	6.9	6.0	5.1	4.9	4.3	3.3	2.4	1.5	0.5	0.0	0.0	0.0	0.1	1.0	1.9	3.0	4.1	5.2	5.3	6.4	7.6	8.7	9.9	10.4	10.5	10.5	
May	10.3	9.5	8.6	7.7	6.8	5.9	5.3	5.0	4.1	3.1	2.1	1.1	0.1	0.0	0.0	0.3	1.3	2.4	3.6	4.7	5.6	5.8	7.0	8.1	9.2	10.2	11.2	11.2	11.2	11.1	10.2
Jun	9.3	8.4	7.5	6.6	5.6	5.6	4.7	3.7	2.7	1.6	0.6	0.0	0.0	0.3	1.4	2.6	3.8	4.9	5.7	6.0	7.1	8.2	9.2	10.2	11.1	11.4	11.4	11.4	10.7	9.8	
Jul	8.8	7.9	7.0	6.0	5.8	5.0	4.0	2.9	1.9	0.8	0.0	0.0	0.1	1.3	2.5	3.6	4.8	5.5	5.8	6.9	7.8	8.8	9.7	10.5	11.0	11.0	11.0	10.8	9.8	8.9	7.9
Aug	6.9	6.0	5.5	4.9	3.9	2.8	1.8	0.8	0.0	0.0	1.0	2.1	3.3	4.4	5.1	5.4	6.3	7.2	8.1	8.8	9.5	10.1	10.1	10.1	10.0	9.5	8.5	7.5	6.5	5.5	
Sep	4.9	4.4	3.4	2.5	1.6	0.7	0.0	0.0	0.0	0.5	1.6	2.7	3.8	4.7	4.7	5.6	6.4	7.1	7.7	8.3	8.8	8.9	8.8	8.8	8.7	7.9	6.8	5.8	4.8	4.1	
Oct	3.8	2.9	2.1	1.3	0.6	0.0	0.0	0.0	0.0	0.9	2.0	2.9	3.7	4.1	4.4	5.1	5.7	6.2	6.7	7.2	7.4	7.3	7.3	7.2	7.1	6.1	5.1	4.2	3.3	3.2	2.5
Nov	1.8	1.1	0.5	0.0	0.0	0.0	0.0	0.8	1.6	2.3	3.0	3.3	3.5	4.0	4.5	4.9	5.4	5.8	5.8	5.7	5.7	5.6	5.6	4.7	3.9	3.2	2.5	2.5	1.9		
Dec	1.3	0.7	0.0	0.0	0.0	0.0	0.0	0.2	0.8	1.4	1.8	2.3	2.6	2.8	3.3	3.8	4.3	4.9	4.9	4.9	4.9	4.9	4.9	4.4	3.8	3.2	2.6	2.5	2.1	1.5	

Index

A

achromats 49–50
adjustment layers 124, 164
Adobe RGB (color space) 103
airmass 91–93
Airy disk 47, 55–56, 69, 94, 147
alignment 108–110
Alnitak 96
alt-azimuth mount 38–39
altitude 91
Analog-to-Digital Converter (ADC) 14, 16, 32–33, 35, 85, 89
Analog-to-Digital Units (ADUs) 14, 32–33, 85
angular resolution 55
anti-blooming gates 22
aperture 46
apochromats 49–50
Arp 65
ASCOM (AStronomy Common Object Model) 70, 73
astigmatism 48, 52, 63, 95
Astro Photography Tool 83
Astro-Physics 61
AstroPlanner 68
AstroSysteme Austria (ASA) 61
AstroTech 61
Atlas of Peculiar Galaxies 65
atmospheric dispersion correctors 92
atmospheric extinction 91
atmospheric refraction 92
atmospheric seeing 58
auto-adaptive weighted average 111
Auto Color tool 144
autoguiding 40, 71–74, 86, 96

B

backlash 74
Backyard EOS 83
baffles 33
Bahtinov mask 70–71
Balmer series 82
Barlow lens 63
Barnard catalog 65
batteries 77
Bayer matrix 21
bayonet mount 44
bias frames 32–33, 88, 89, 96, 107
bias signal 31, 32, 34, 35, 88
binning 28, 43, 61
bit depth 16–18, 32
    and posterization 133
black-body radiation 104
black point 122
blending modes (layers) 125–126
    combining with high pass filter 152
Bode's Galaxy 67
Borg 54, 61
Brightness/Contrast tool 138
bulb 83, 86

C

Caldwell catalog 65
calibration 32, 34, 84, 87, 107–116
California Nebula 66
Canon 43, 60
    raw files 100
Carey mask 70
Cartes du Ciel 68
Cassegrain telescope design 52
catadioptric telescope design 52
Catalog of Bright Diffuse Galactic Nebulae 65
CCD 15–16, 21–22
CCDSoft 83, 89
CCDStack 119, 148
Ced 214 67

Cederblad catalog 65
Celestron EdgeHD 61
centroid calculation 74
CFHT (Canada France Hawaii Telescope) palette 142
chromatic aberration 47, 63, 95
chrominance noise 144, 162
  and sharpening 149–151
clamshell rings 77
cleardarksky.com 85
clipping 17, 84, 122, 129
clipping layer mask 125
  for color synthesis 143
CMOS 15–16, 21–22
CMYK color system 102
coefficient of variation 23
cold pixels 31, 33, 107
  correction in calibration 111, 113
color 166
  balancing 130, 144
  effect of stretching on 134
  saturation 133, 134, 144–145
  synthesis 140–142
color accuracy 105
Color Balance tool 144
color blending mode 126, 144, 161
color burn blending mode 125
color calibration 105–106
colorimeter 103
color management 101, 103
Color Merge 140
color profiles 103
Color Range tool 127, 136–138, 160
  for star selection 157
Color Sampler tool 130
color spaces 101, 103–104
color temperature 104
color vision 100–101
column defects 111
coma 47, 51, 61
Coma cluster 68
composite layer 155, 163
composite layers 126–127, 137, 151
composition 86
cone cells 101
content-aware tools 138

contrast
  creating via S-curves 123
  local enhancement 147–154
  selective adjustments 105
convolution 147
.CR2 files 100
Crab Nebula 67
Crayford focusers 71
crown glass 49
.CRW files 100
curves tool 121–124, 129–133
  anchors 131
  for color correction 136
  for microcurves 154
  for removing objects 137

D

dark current. *See* thermal signal
darken blending mode 125, 139, 156
darker color blending mode 125
dark frames 31–32, 87–88, 96, 107
Dawes' Limit 56
decibel (dB) 17
declination 39
deconvolution 147
Deep-Sky Planner 68
DeepSkyStacker 111, 119
  auto-adaptive weighted average 111
  background calibration 111
  entropy weighted average 111
  groups 112
  quality scores 113
  reference frame 112
DeNoise 162
desaturation
  of background noise 162
  of stars for selection 157
Despeckle filter 162
dew 95, 116
dew heaters 78
dew prevention 78
dew shields 78
difference blending mode 126, 138
  to create a star mask 156
diffraction 47, 55–56
diffraction-limited optics 47

Index

diffraction patterns 94–95
Digital Development Process 135
distortion 48
dithering 89
  between exposures 115
divide blending mode 126
.DNG files 100
Dobsonian telescopes 51
30 Doradus 67
dovetail plate 77
Draper, Henry 67
Dreyer, John 65
drift alignment 40, 42–43
drizzle algorithm 115–116
DSLR (Digital Single-Lens Reflex) 43, 44, 61, 83, 85, 87, 88
  long exposure noise reduction 111
  raw files 100
Dumbell Nebula 67
dust 33, 89
  in calibrated images 116
Dust and Scratches filter 137, 155
DynamicBackgroundExtraction tool 136
dynamic range 17, 84, 86, 100
  contents of in astroimages 120
  reallocation of with curves 121–123
  reallocation of with levels 123

E

Eagle Nebula 66
ED (Extra-low Dispersion) glass 49–50
Elephant's Trunk Nebula 66, 150, 165
emission lines 79–82
entropy weighted average 111
EOS Utility 83
equatorial mount. *See* German equatorial mount
Eta Carina Nebula 66
eXcalibrator (software) 106
exclusion blending mode 126
Exif (metadata) 87, 104
  temperature 110
exit pupil 47
exposure duration 84–85, 86

F

Fastar 53, 61

feathering selections 127–128, 133
  of stars 157
field curvature 48, 50, 61, 95
field de-rotators 38
field flattener 45, 48, 50, 61
field of view 46, 60, 86
field rotation 38–39, 40
file formats 100
film 15, 19
FITS 100
fixed-pattern noise 89
Flame Nebula 67, 96
flat dark frames 107
flat darks 88
flat frame
  gradients in 116
flat frames 33, 86, 88–89, 96, 107
flexure 72
flint glass 49
focal length 46, 60, 79
focal ratio 29, 46, 69, 80, 85
focal reducer 44, 45, 61–63
focus 69, 86, 94, 95, 96
focusers 70
focusing masks 70–71
FocusMax 70
Fornax Nebula 67
Frame-transfer CCDs 22
f-ratio. *See* focal ratio
Fraunhofer, Joseph von 39
full well capacity. *See* well capacity
FWHM (Full Width at Half Maximum) 113

G

G2V stars 105–106
gain 16–18, 85, 86
galaxies
  orientation of 166
galaxy season 68
gamma 19
gamma correction 120
  via levels tool 123
gamut 101, 103
GEM. *See* German equatorial mount
German equatorial mount 38–40, 41, 85
German equatorial mount, balancing 86

197

GIMP (GNU Image Manipulation Program) 119
globular clusters 90, 159
GPUSB 72, 78
gradients 116
   removing 136–138
GradientXTerminator 136
GREYCstoration 162

H

halos 95, 149
   from narrowband color synthesis 160
   from sharpening 147
hard light blending mode 126
hard mix blending mode 126
Hartmann mask 70
HDR Toning tool 154
Heart Nebula 67
Helix Nebula 67
high pass filter 151–152
histogram 19–20, 81, 84
histograms 118–121
Horsehead Nebula 67, 96
HoTech 61
hot pixels 31, 33, 88, 107
   correction in calibration 111, 113
   streaking 116
HSL color model 104
HST (Hubble Space Telescope) palette 142–143, 160
Hubble palette 81
hue blending mode 126, 144
Hue/Saturation tool 133, 136, 143, 144, 160, 164
Hydrogen-alpha emission line 80–82, 85, 142, 160
Hydrogen-beta emission line 80–82, 83, 141
Hyperstar 53, 61

I

IC 353 67
IC 405 142
IC 410 142
IC 434 67
IC 1396 66, 150, 153, 165
IC 1831 67
IC 1848 67
IC 2118 66
IC 2177 155–156
IC 2602 67
IC 4177 127
IC 4592 66
IC 4601 66
IC 4628 67
IC 5067 67
IC 5070 67
image scale 29, 46, 56–58, 162
Images Plus 83
ImagesPlus 148, 162
Index Catalogues (IC) 65–66
interference filters 79
interline CCDs 22
IR filters 43, 85, 95
IRIS 119, 136, 148
   noise reduction tools 162
ISO 85. *See* gain

J

JPEG 100

K

KAF-8300 60
KAI-2020 60
KAI-11000 60
kappa-sigma clipping
   sigma clipping. *See* bias frames
kernel (for convolution) 147

L

LAB (color space) 103–104, 140, 141
   for boosting satuation 145–147
   for sharpening 147
Lagoon Nebula 67
Large Magellanic Cloud 66, 67
lasso (selection tool) 127
   polygonal 138
layer masks 126
   for selective noise reduction 163
   for selective sharpening 151
layers 124–126
   adjustment layers 124
   blending modes 125
   clipping mask 125
   groups 125
lens spacers 94

Index

levels tool 123, 129
  for adjusting layer masks 151
  for color saturation 145
lighten blending mode 144
light frames 31, 107
light pollution 85, 90
light pollution filters 82, 90, 96
linear burn blending mode 125
linear light blending mode 126
Liquify tool 160
local contrast enhancement 147–154
Lord mask 71
Losmandy-style dovetail 77
LRGB filters 79
LRGB images 140
luminance exposures 140–141
luminance noise 162
luminosity blending mode 126, 140, 147
luminosity class (stellar) 106
Lynd's Catalogue of Bright Nebulae (LBN) 65

M

M1 67
M4 66
M8 67
M13 67, 90
M16 66
M27 64, 67
M31 64, 66
M33 64, 67
M42 66, 67
M42 (threaded fitting) 44
M43 66
M45 66
M51 64, 67, 130
M57 64, 68
M63 67
M81 67
M86 137
M101 28–29, 67
M104 68
M106 111–113
magnitude, visual 66
Maksutov-Cassegrain telescopes 53, 94
Maksutov-Newtonian telescopes 53
Markarian's Chain 66

Match Color tool 144
Maxim DL 73
MaximDL 83, 87, 89, 119, 148, 162
median stacking 110
meridian flip 39, 42, 85, 86
Merope Nebula 67
Messier catalog 65
Messier, Charles 65
metadata 100
  Exif 100
  in raw files 103
microcurves 154
microlens 44, 95
microlenses 22
Minimum filter
  for star reduction 158–159
min-max stacking 111
mirror flop 53, 94
moonlight 80, 83, 85, 90
mounts 38–39
  balancing 94
move tool 139
multiply blending mode 125

N

narrowband filters 20, 79–82, 90, 96
  color synthesis 141
narrowband imaging 85, 105
  color synthesis 141
  combining with RGB data 144
  star color 160
Neat Image 162
Nebulosity
  noise reduction tools 162
Nebulosity (software) 83, 89, 119
.NEF files 100
New General Catalogue (NGC) 65–66
Newtonian telescopes 51, 94
NGC 55 67
NGC 104 67
NGC 281 120
NGC 598 67
NGC 896 67
NGC 1360 67
NGC 1435 67
NGC 1499 66, 165

NGC 1909 66
NGC 1952 67
NGC 1976 67
NGC 2024 67
NGC 2070 67
NGC 3031 67
NGC 3372 66
NGC 4594 68
NGC 5055 67
NGC 5139 67
NGC 5194 67
NGC 5195 132
NGC 5457 67
NGC 6523 67
NGC 6526 67
NGC 6530 67
NGC 6611 66
NGC 6720 68
NGC 6726 67
NGC 6727 67
NGC 6853 67
NGC 7000 66, 152
NGC 7293 67
NGC 7822 67
Nikon 43, 60
Nitrogen-II (NII) emission line 80–82, 141
noise floor 123
Noise Ninja 162
noise reduction 162–163
　applying selectively 163
North America Nebula 66
North America Nebula, 67

O

oblong stars 139
off-axis guiding 72
offset filter 139
offset frames. *See* bias frames
offset signal. *See* bias signal
Okano, Kunihiko 135
Omega Centauri 67
one-shot color
　for star color 144
one-shot color (OSC) 20–21, 79, 82, 140
　raw files 103
open clusters 90

orientation 165
Orion (manufacturer) 60
Orion Nebula 66, 67
overlay blending mode 126, 152
oversampling 56–58, 59
Oxygen-III (OIII) emission line 80–82, 83, 142, 160

P

Pac-Man Nebula 120
parfocal filters 80
Patch tool 138
PEC. *See* Periodic Error Correction
.PEF files 100
Pelican Nebula 67
Pentax 43, 60
　raw files 100
periodic error 41–42
Periodic Error Correction 41–42
Petzval curvature 48
Petzval telescope design 50
PHD (software) 73, 89
photometry 79
Photoshop 119
　versions 118
photosite 14
pier mount 76
pin light blending mode 126
Pinwheel Galaxy 67
Pixel 14
PixInsight 119, 128, 136, 148
　noise reduction tools 162
planetarium software 68
Pleiades 66
Pogson, Norman 66
point spread function 147–148
Poisson distribution. *See* Poisson process
Poisson process 24–26
polar alignment 40–42, 71, 76, 86
　and oblong stars 139
polar alignment scope 40
posterization 110, 133–134
　from Dust and Scratches filter 155
power supply 78
Prawn Nebula 67
prime lens 50

Q

QSI 60
quantization error. *See* quantization noise
quantization noise 23, 32–33, 34
quantum efficiency 14–16

R

rack-and-pinion focusers 71
raw image format 86, 100
Rayleigh criterion 55
Rayleigh scattering 80
RCW catalog 65
read noise 16–18, 23, 32–33, 34, 84, 88, 89
Rector, Travis 143
Reduce Noise filter
  Reduce Color Noise 160, 162
Registax 119
resampling 165
resolution 55–56
Revised New General Catalogue and Index Catalogue 65
RGB color system 102–103
Rho Ophiuchus 66
Richardson-Lucy algorithm 148
right ascension 39
ringing artifacts 147, 149
Ring Nebula 68
Ritchey-Chrétien telescopes 48, 52, 61, 71, 94
Rosette Nebula 67, 165
rule of thirds 165

S

saddle 77
Sadr 151
saturation blending mode 126, 146
SBIG (Santa Barbara Instruments Group) 60, 72
Schmidt-Cassegrain telescopes 52–54, 61, 78, 94
Schmidt-Newtonian telescopes 53
screen blending mode 126, 143, 156
Sculptor Galaxy 67
S-curves 123
seeing. *See* atmospheric seeing
seeing (atmospheric) 90
selection tools 127–128
Selective Color tool 102, 142, 144, 160

sensel 14
sensor sizes 44
serial adapters 78
servo motors 70
Sh2-170 67
Sh2-171 67
Sh2-190 67
Sh2-199 67
Shannon-Nyquist sampling theorem 56–58
sharpening 147–154
Sharpless catalog (Sh2) 65
Shoestring Astronomy 73, 78
shot noise 23, 24–26, 26–29, 34
sigma clipping 84, 111, 138
signal-to-noise ratio 23, 26–29, 34–36, 79, 80, 84, 96, 128
  and image scale 162
  and luminance data 141
  and sharpening 147
skyglow 23–24, 29–30, 34, 82, 84, 90
Sky Tools 68
Small Magellanic Cloud 66
SNR. *See* signal-to-noise ratio
soft light blending mode 126, 152
Sombrero Galaxy 68
Sony 43
Soul Nebula 67
Southern Pleiades 67
Sparrow's limit 56
spatial resolution 55
spectral class (stellar) 106
spherical aberration 48
Spherize filter 159
spherochromatism 48
Sponge tool 146
Spot Healing Brush 138
sRGB (color space) 103
ST-4 (autoguiding port) 72, 78
stacking
  statistical methods for 110–111
stamp layers. *See* composite layers
standard deviation 25–26
Stark, Craig 73
star mask 155, 159
Starry Night (software) 68
stars

color in narrowband images 143
color saturation 146
removal 155
Steinicke, Wolfgang 65
Stellarium 68
stepper motors 70
stretching 129–133
linear vs. non-linear 129
subtract blending mode 126
Sulfur-II (SII) emission line 80–82, 142, 160
Sunflower Galaxy 67
synthetic flats 137
synthetic luminance 141

T

T-adapter 44
Takahashi 54, 61
Tarantula Nebula 67
telephoto lens 50–51, 66
Televue 54, 61
Tempel's Nebula 67
temperature 93
thermal signal 23–24, 30–32, 34, 77, 87
in dark frames 107
thermo-electric cooling (TEC) 87
TheSky (software) 68
TIFF 100
transparency (atmospheric) 90
Triangulum Galaxy 67
T-ring 44
T-thread 44
47 Tucanae 67

U

UBVRI filters 79
undersampling 57–58
unsharp mask 148
for local contrast 148–151
Uppsala General Catalogue of Galaxies (UGC) 65
USB hubs 73, 76, 78

V

van den Bergh catalog (vdB) 65
Veil Nebula 66
vignetting 33–34, 89, 136
Virgo cluster 66, 68, 137

Vixen-style dovetail 77

W

wavelength-ordered palette. *See* HST (Hubble Space Telescope) palette
well capacity 16–18, 84, 85
Whirlpool Galaxy 67, 130
white balance 85, 104
white point 122
William Optics 61
wind 94, 139
Witch Head Nebula 66, 90
Wynne corrector 52, 61

Z

zodiacal light 90